Harvey Washington Wiley

Record of Experiments at Fort Scott, Kansas

In the manufacture of sugar from sorghum and sugar-canes, in 1886

Harvey Washington Wiley

Record of Experiments at Fort Scott, Kansas
In the manufacture of sugar from sorghum and sugar-canes, in 1886

ISBN/EAN: 9783337130107

Printed in Europe, USA, Canada, Australia, Japan

Cover: Foto ©berggeist007 / pixelio.de

More available books at **www.hansebooks.com**

U. S. DEPAR. MENT OF AGRICULTURE.

DIVISION OF CHEMISTRY.

BULLETIN   No. 14.

RECORD OF EXPERIMENTS

AT

FORT SCOTT, KANSAS,

IN

THE MANUFACTURE OF SUGAR

FROM

SORGHUM AND SUGAR-CANES,

IN

1886.

BY

H. W. WILEY,

CHEMIST.

WASHINGTON:
GOVERNMENT PRINTING OFFICE.
1887.

UNITED STATES DEPARTMENT OF AGRICULTURE,
DIVISION OF CHEMISTRY,
Washington, D. C., December 21, 1886.

SIR: I beg leave to submit herewith a report of the work done at Fort Scott during the present year under authority of Congress in "Experiments in the manufacture of sugar from sorghum and sugar-cane by the processes of carbonatation and saturation."

The conduct of this work you placed in my hands, and throughout the whole of it I have had your earnest support.

The results of the work are now presented for your inspection and approval.

Very respectfully,

H. W. WILEY,
Chemist.

Hon. NORMAN J. COLMAN,
Commissioner of Agriculture.

3

EXPERIMENTS IN THE MANUFACTURE OF SUGAR FROM SORGHUM.

The results of the experiments made at Ottawa last year gave encouragement to the friends of the sorghum sugar industry, and led to the undertaking of a new series of experiments at Fort Scott.

The Department of Agriculture entered into the following agreement with the Parkinson Sugar Company at Fort Scott:

WASHINGTON, D. C., August 7, 1886.

AGREEMENT BETWEEN THE COMMISSIONER OF AGRICULTURE AND THE PARKINSON SUGAR COMPANY OF FORT SCOTT, KANS.

The Commissioner of Agriculture agrees to erect at the works of the Parkinson Sugar Company of Fort Scott, Kans., one diffusion battery with all its appliances; three cane-cutters, one of which shall have a horizontal cutting disk, with appliances for feeding the cane to the same, and elevators for delivering the chips to the cells.

He further agrees to erect one carbonatation apparatus, to consist of a lime-kiln, carbonic-acid pump, four carbonatation tanks, and four filter-presses, with all their connections; also one sulphur apparatus, consisting of two sulphur furnaces, three saturation-tanks, three filter-presses, one air-pump, and all necessary connections.

He further agrees to prepare the whole of the above-mentioned machinery for practical work, and to provide all necessary labor and material for a thorough experimental trial of the same, and when this trial is finished to allow the Parkinson Sugar Company the free use of the apparatus for the rest of the manufacturing season of 1886, without any charge for rental to the Parkinson Company aforesaid.

It is expressly agreed and understood that all machinery furnished by the Department of Agriculture, and all fixtures and appliances therewith connected, shall remain the property of the Department, and the Commissioner reserves the right to make such disposition of all of it after the end of the present manufacturing season as may seem to him best suited to promote the public interest.

The Parkinson Sugar Company agree to furnish suitable buildings in which to erect this machinery, to supply steam for driving it and for use in the calorisators of the battery, and to allow the Commissioner of Agriculture as much time as he may desire, not exceeding ten days from the commencement of the manufacturing season, for the purpose of making the experimental trials before mentioned; provided that during these experimental trials the Commissioner of Agriculture shall pay for all coal consumed for supplying the steam mentioned above, and for all limestone, coke, sulphur, filtering-cloths, and other materials used in the experiments.

The said company also agree to furnish a suitable room for the chemical laboratory to be erected by the Department and used by the Department chemists during the continuance of the manufacturing season.

It is further agreed on the part of the said Parkinson Company that during the period of the experiments mentioned the accredited representative of the Department at Fort Scott, namely, the chemist of the Department, or such other person as the

5

Commissioner may designate, shall have sole control and direction of the work, in so far as the extraction and purification of the sugar-juices are concerned.

Further, on the part of the Commissioner of Agriculture, it is agreed that during the entire manufacturing season he will supply the services of one superintendent, namely, Prof. M. Swenson, and one sugar-engineer, namely, Mr. G. L. Spencer, or some other persons of equal experience and ability, and also a competent corps of chemists; provided the company aforesaid give to said agents of the Department every facility for studying the processes employed, and supply them with full and accurate data of the amount of cane entering into manufacture, the quantities of sugar and sirup made, and all other information which will help the Commissioner to make a full and accurate report of the whole work; provided further, that after the experimental work above mentioned has been finished and during the time the said company operate the machinery for the purpose of manufacturing sugar and sirup for profit, the Department of Agriculture shall not be responsible for any other expenses than those which relate to the employment of the agents of the Department above mentioned.

NORMAN J. COLMAN,
Commissioner of Agriculture.
PARKINSON SUGAR COMPANY,
By C. F. DRAKE, *President.*

The Congress having made an appropriation of $94,000 for the continuance of the experiments, the following contract was made between the Commissioner of Agriculture and The Pusey & Jones Manufacturing Company of Wilmington, Del., for the construction and erection of the necessary machinery.

WASHINGTON, D. C., *April 21, 1886.*

DEAR SIR: I desire to secure, for the experimental sugar station which the Department will establish in connection with the Parkinson Sugar Company, at Fort Scott, Kans., a diffusion battery. Will you kindly send me estimates of the cost of the battery, in conformity with the following general requirements?

(1) The battery to be of a capacity to work 200 tons of cane in twenty-four hours at a mean rate.

(2) The battery to consist of fourteen cells, arranged in a straight line, with valves, calorisators, and connections complete.

(3) The cells to be cylindrical, and have a discharge-gate at the bottom of the area of the cross section of the cell.

(4) The valves to be so arranged that the water can be introduced at top or bottom of each cell at the pleasure of the operator.

(5) The joint of the discharge-gate to be made by hydraulic closure.

(6) The last charge of water in each cell to be removed by compressed air.

(7) Apparatus for the automatic charging of the cells with fresh chips.

(8) Apparatus for removing the exhausted chips.

(9) Calorisators to be furnished with thermometers, with face like steam-gauge.

(10) Measuring tanks for withdrawing juice, with accurate float-gauge.

(11) Two cane-cutters, with vertical disks, and forced feed, with cane-carriers and chip-elevators complete; these to be simply those already at Ottawa, with a modification of the forced feed, to prevent choking.

(12) Air compressor and reservoir for discharging water from cell next to be emptied.

In the above apparatus all the valves, piping, shafting, pulleys, elevators, &c., which were used at Ottawa are to be incorporated in the new machinery where it is possible without disadvantage, and to be valued at their original cost price.

In your proposals, which I hereby ask for, please give all the details of the apparatus which must be guaranteed to work and give satisfaction to the Department.

Since the proper erection of this machinery is also essential to its success, I will ask you to submit a proposal to erect said machinery at Fort Scott and deliver it to the Department in proper working order on or before the 10th of August, 1886
Respectfully,

NORMAN J. COLMAN,
Commissioner.

WM. G. GIBBONS,
President, &c., Wilmington, Del.

———

WILMINGTON, DEL., *May* 8, 1886.
DEAR SIR: Replying to your favor of 21st ultimo, received three days ago, we offer to build the machinery therein specified, to say—

A diffusion battery, consisting of 14 cells, cylindrical in form, 44 inches in diameter, 7 feet 4 inches long, with door at bottom of full diameter of cell, and having counterbalance and hydraulic-joint packing; valves arranged so that the water can be introduced into cells at either top or bottom at pleasure.

An air-compressor and reservoir so arranged that the water in each cell can be removed by compressed air; apparatus for automatic charging of the cells with fresh chips and removing the exhausted chips to a comfortable distance from the battery.

Calorisators to be furnished with thermometers. Unfortunately those made in this country with face like steam-gauges are so slow of operation, that they would be useless. We are forced, then, to supply mercurial thermometers; will select the plainest dials to be had.

Proper measuring-tanks for withdrawing juice with floating gauge.

Alter the two cane-cutters now at Ottawa, Kans., so that the forced feed shall not choke, and supply cane carriers and chip-elevators. Price, $14,125.

In this it is proposed to use such portions of the valves, pipes, and other things pertaining to the apparatus at Ottawa built by us as may be adaptable to the above.

We also propose to transport all of the above to Fort Scott, Kans., and erect at the works of the Parkinson Sugar Company and have in operation on or before the 10th day of August, 1886, for the further sum of $2,500.

Soliciting the order, which shall have prompt dispatch, we are,
Yours, truly,

THE PUSEY & JONES COMPANY,
By WILLIAM G. GIBBONS,
President.

Hon. NORMAN J. COLMAN,
Commissioner of Agriculture, Washington, D. C.

———

WASHINGTON, D. C., *July* 26, 1886.
GENTLEMEN: I have received your communication of 25th instant in respect of the amount which you offer us in exchange for the machinery specified in my letter of 22d instant, and your offer is satisfactory to me. I therefore accept your proposition of 8th of May, last, viz:

"A diffusion battery consisting of fourteen cells, cylindrical in form, 44 inches diameter, 7 feet 4 inches long, with door at bottom of full diameter of cell, and having counterbalance and hydraulic joint packing; valves arranged so that the water can be introduced into the cells at either top or bottom at pleasure.

"An air-compressor and reservoir, so arranged that the water in each cell can be removed by compressed air; apparatus for automatic charging of the cells with fresh chips and removing the exhausted chips to a comfortable distance from the battery.

"Calorisators to be furnished with thermometers. Unfortunately those made in this country, with face like steam gauges, are so slow of operation that they would be useless. We are forced, then, to supply mercurial thermometers. Will select the plainest dial to be had.

"Proper measuring tanks for withdrawing juice, with floating gauge.

"Alter the two cane-cutters now at Ottawa, Kans., so that the forced feed will not choke, and supply cane-carriers and chips elevators. Price, $14,125.

"In this it is proposed to use such portions of the valves, pipes, and other things pertaining to the apparatus at Ottawa built by us as may be adaptable to the above.

"We also propose to transport all of the above to Fort Scott, Kans., and erect at the works of the Parkinson Sugar Company, and have in operation on or before the 10th day of August, 1886, for the further sum of $2,500.

Replying further to your letter of 25th instant, I will say that the cane-cutters and battery now at the "Hermitage" plantation of Mr. D. F. Kenner, inLouisiana, will be delivered alongside the Cromwell Wharf, in New Orleans, before the 1st of September next, in accordance with your desires.

In further preparation of the work at Fort Scott, I desire you to submit to me your estimates of the cost of four filter presses and a sufficient number of carbonatation tanks, to be used in the experiments in the manufacture of sugar at Fort Scott during the coming campaign.

I desire this proposition to include the freight to Fort Scott; in other words, I ask you to deliver the apparatus just mentioned to the Department at Fort Scott, Kans., at the earliest possible moment.

Very respectfully,

NORMAN J. COLMAN,
Commissioner.

THE PUSEY & JONES COMPANY,
Wilmington, Del.

A contract was also made for a part of the apparatus for treating the diffusion juice with lime and carbonic acid in the following terms:

WILMINGTON, DEL., *August* 3, 1886.

DEAR SIR: We owe you an apology for so much time having been allowed to elapse since the receipt of your favor of 26th ultimo, and its reply. Illness in the family of the writer has prevented his attention, and hence the delay, which please excuse.

The diffusion machinery referred to in your letter is now being erected at Fort Scott, Kans., at the works of the Parkinson Sugar Company. Of the date of its starting we shall advise you later.

The four filter presses you inquire for will cost, delivered at Fort Scott, complete, all allready for service, $1,100 each. Four carbonatation tanks, each 6 feet 6 inches long, 4 feet 6 inches wide, and 6 feet 6inches high at front, and 6 feet high at back, with receiving and discharge pipe and valves, gas-pipe, and distribution, copper coil heater, and vapor pipe, all complete, delivered at Fort Scott, Kans., $350 each.

Soliciting your order, we are yours, truly,

THE PUSEY & JONES COMPANY,
By WM. G. GIBBONS, *President.*

Hon. NORMAN J. COLMAN,
Commissioner of Agriculture, Washington, D. C.

The battery erected by the Pusey & Jones Company, consisted of 14 cells, arranged in single line, with calorisators and apparatus for use of compressed air in discharging the water from each cell before dropping the exhausted chips. The working of the battery was entirely satisfactory.

Each cell had a capacity of 75 cubic feet, and would hold 1,900 pounds of sorghum chips, moderately packed. Each cell was constructed from the drawings obtained from the Fives-Lille Company, and the detailed description may be found in Bulletin No. 8.

The cutters used were those employed at Ottawa last year. The contractors made no attempt whatever to rebuild the forced feed attachment, and this failure was the cause of the chief delay we experienced after the apparatus was in regular use. With very sharp knives, and with cane fresh and green, they did reasonably good work, but after a frost had killed the leaves of the cane it was found almost impossible to make the cutters work. It often required half an hour to fill a single cell. When it is remembered that the rest of the apparatus could easily have worked a ton of chips each eight minutes, the disastrous effects of this delay can be appreciated.

From this cause great trouble was experienced in working the battery. When all the cells were in use each one was often under pressure three or four hours. The cane was unusually acid, and from this there followed a large inversion of sucrose in the battery. If, to avoid this, the temperature of diffusion was lowered, fermentation would set in. There was nothing left for us to do but to work a smaller number of cells. Often only six or seven cells were under pressure, and consequently the degree of extraction was far less perfect than it would have been otherwise.

The style of cutter used furnished a chip well suited to diffusion, but I am convinced that these cutters are more costly and require more power for operation than is necessary.

With a view of correcting these defects I purchased a beet-root cutter, formerly used by the Portland Beet Sugar Company, and had it rebuilt by the Colwell Iron Company of New York, for an experimental cane cutter.

This apparatus had a horizontal disk, and was so modified as to take a multiple feed, the cane being delivered to it through six hoppers inclined 40 degrees to the vertical. With perfectly clean canes this cutter gave promise of success, but with the sorghum-cane as it came from the field it proved a total failure.

This leads me to believe that the cutters used at Java and other places so successfully with sugar-cane would not serve the purpose of slicing sorghum for the battery. Any question of cleaning the canes before delivering them to the cutter must be negatived on the score of economy.

For the further study of the problem I tried the system of cane-slicing invented by Mr. H. A. Hughes, of Rio Grande, N. J.

The principle of this system consists in first cutting the canes into lengths of three or four inches by means of an ensilage-cutter, and after passing them through a cleaning apparatus deliver them to a shaving-machine constructed on the principle of a board-planer.

This latter part of the apparatus was kindly loaned to the Department by Mr. Hughes.

The canes were first cut by a Belle City ensilage-cutter into pieces about 2.25 inches in length. These pieces were run through a fanning-mill and nearly all the blades and sheaths were thus removed. The clean pieces of cane were next delivered to a slicer built on the principle of an ordinary board-planer. The cylinder was 6 inches in diameter and 30 inches in length, and carried two knives projecting one-eighth to one-sixteenth inch beyond the surface. This was driven at a high rate of speed, over 3,000 revolutions per minute. The canes were shredded rather than sliced by this process, so that the extraction of the sugar was rather a maceration than a diffusion.

Even with this small machine it was found possible to prepare nearly as much cane for the battery as with the three ponderous cutters described. It was found, however, that the ensilage-cutter was not strong enough to do the work, and hence this most promising system of cane-cutting, practiced successfully at Rio Grande, was discontinued. The experiment, however, led me to believe that the principle was the right one; especially is this so because it permits of the easy cleaning of the canes by first cutting them into small pieces. This seems to be the only practical way of accomplishing what is of prime necessity to diffusion, viz, the removal of all deleterious substances from the chips.

Having demonstrated the practicability of cleaning the cane in the manner already described, my attention was next directed to the consideration of the best method of cutting the short pieces of cane into chips suitable for diffusion. For this purpose I had constructed by the Fort Scott Foundry a centrifugal slicer. The theory of this apparatus was that the knives, being carried in a revolving frustum of a cone, and the short pieces of cane being fed from the inside of this cone, the chips, as soon as cut, would fly off by centrifugal force. A trial of this apparatus showed that the fiber of the cane would clog the knives and thus stop the work. The close of the season prevented any modification of the apparatus. I think the principle of the apparatus is promising enough to warrant further trial.

As a result of the experiments with cutters the following conclusions can be drawn:

(1) Whatever the form of the cutting-machine employed may be, it is necessary that the cane be cleaned. This cleaning should not consist of the removal of the blades alone, but also the sheaths.

(2) The slicing of the canes obliquely by means of a vertical cutting-machine with a forced feed is not an economical method of procedure.

(3) The use of a cutting-machine with a horizontal disk and multiple feed is impracticable for sorghum canes unless they are perfectly clean.

(4) The preliminary cutting of the canes into short lengths promises the easiest solution of the problem of cleaning the cane.

(5) The subsequent slicing of these sections by some form of apparatus is a mechanical problem which can be solved.

THE APPARATUS FOR DELIVERING THE CHIPS TO THE BATTERY AND REMOVING THEM THEREFROM.

The working of the chip elevators and the apparatus .or removing the exhausted chips was exceedingly unsatisfactory.

The chips falling into the pit below the cutters were carried by a screw conveyor to a bucket elevator. Thence they were dropped onto a belt conveyor, which delivered them to the apparatus for blowing out the leaves, &c. The screw, the elevator, and the belt frequently became choked and occasioned a great deal of trouble and delay.

The apparatus for removing the exhausted chips gave still greater trouble.

In discharging a cell the whole contents, weighing a ton, were thrown at once on the conveyor. This load was too great, and many days' delay were experienced in making the alterations necessary even to moderate efficiency.

The elevator for taking the exhausted chips from this conveyor was a very complicated and inefficient piece of apparatus, and many tedious changes had to be made before it would do the necessary work. Finally its use was abandoned altogether. The lessons taught by these unfortunate delays show that the proper method for removing the exhausted chips from the battery is by means of a tramway and dumpcart, as practiced at Almeria and described in Bulletin No. 8. A great deal of apparatus and power will be saved by this method of disposing of the chips. The conveyor for filling the cells worked in marked contrast with the rest of the chip-handling machinery, and gave perfect satisfaction. This conveyor extended the entire length of the battery, and was placed directly above it. Over each cell was a door in the floor of the conveyor. When a cell was to be filled the door above it was opened and the chips fell through onto a funnel which directed them into the cell. The bottom of the conveyor at Fort Scott was too near the top of the cells. It should be not less than 6 feet above the top of the cells, so as to allow ample room for tamping the chips as they fall into the cell, thereby greatly increasing the capacity of the battery. I do not think a better contrivance could be devised for filling the cells of a line battery. I am still of the opinion, however, that the charging of a circular battery, as described in Bulletin No. 8, would be a more simple method. The disposition of the battery, however, is not a matter of vital importance.

I am further of the opinion that it will not be difficult for an ingenious mechanical engineer familiar with elevating apparatus to build the machinery which will elevate the cuttings to the battery without any difficulty. By the employment of the centrifugal cutter already described, which can be placed directly over the battery, the elevators will only have to carry the short pieces of cane, a very easy task.

MACHINERY FOR HANDLING THE CANE.

The apparatus for taking the cane from the carts and delivering it to the cutters was designed by Mr. W. L. Parkinson. The carts for bringing the cane from the fields are provided with a rack of peculiar construction. On this rack are placed ropes in such a manner that when the cart arrives at the unloading station the ropes can be brought together, inclosing the whole load of cane. By means of a power drum the entire load is drawn from the cart onto a weighing-truck running on a tramway.

As soon as the weighing is completed the truck is moved along the way until it comes opposite the cane-carrier. ' It is drawn from the truck by means of a power drum and is dragged down an inclined plane in large armfuls to the carrier. The carrier runs at right angles to the length of the cane and to the elevators which deliver the canes to the cutters. As the cane is carried along this feed-table the heads are cut off by a circular saw running at a high rate of speed. The heads which escape the saw are afterwards cut off by hand. The canes then pass to a point midway over the three elevators leading to the cutters. Thence by means of an ingenious contrivance it can be dropped into either carrier at will. The apparatus worked well, but aside from the removal of the tops I doubt whether so complicated a piece of machinery is necessary.

CARBONATATION APPARATUS.

This apparatus consists of a lime-kiln, washer for the gas, carbonic-acid pump, and carbonatation tanks.

LIME-KILN.

The lime-kiln was built by Mr. G. L. Spencer, with castings and plans from the Hallesche Maschinenfabrik. The pump was built by the same firm, but was purchased, as well as the castings just mentioned, from the Portland Beet Sugar Company. After the workmen learned how to conduct the operations at the kiln we had no trouble with its manipulation. It furnished an abundant supply of gas, and an amount of lime in large excess of the quantity required.

The limestone at first furnished contained a large quantity of cement and was unfit for use. In all, several days' delay was caused by this imperfection.

After reasonably good limestone was obtained all worked well. The analyses of the limestones employed will be found among the analytical data. The drawings and detailed description of the lime-kiln are found in Bulletin No. 8.

THE PUMP.

The pump was delivered to us in that state of imperfection which three months of very hard usage and six years of disuse produce.

Nevertheless, after a proper adjustment it worked with perfect satisfaction. In all not more than half a day's delay was caused by the adjustment of this apparatus.

THE CARBONATATION TANKS.

These tanks were built by the Pusey and Jones Company, according to the drawings and specifications in Bulletin No. 8, and gave perfect satisfaction. I can suggest no improvement in them unless it be the insertion of revolving paddles to keep down the foam.

THE FILTER-PRESSES.

These, four in number, and of thirty chambers each, were constructed by the Pusey and Jones Company, on the general plan of the Kroog filter-press, but with certain modifications suggested and patented by Mr. Swenson. Their work gave perfect satisfaction. The only fault discovered in them was the weakness of the plates, a great number of them breaking under the ordinary pressure.

THE SULPHUR APPARATUS.

This apparatus consists of an air-compressor, two sulphur furnaces, three sulphuring-tanks, and three Kroog's twin filter-presses. The whole apparatus was built by the Sangerhauser Maschinenfabrik, and its work gave entire satisfaction. The apparatus is described in detail in Bulletin No. 8.

The whole of the machinery, with the unimportant changes noted, was constructed according to the drawings and specifications printed in Bulletin No. 8. Their reproduction is not considered necessary here.

ANALYTICAL DATA.

The analyses of canes, chips, waste-waters, purified juices, &c., were made at the factory chiefly by Dr. C. A. Crampton, assisted by Mr. N. J. Fake. The limestones, masse-cuites, press-cakes, &c., were examined in the laboratory at Washington.

The analyses of the gases from the lime-kiln were made by Mr. G. L. Spencer.

Limestones, &c.

Serial No.	No.	Water.	Carbonic acid CO₂.	Insoluble and silica, SiO₂.	Fe₂O₃, Al₂O₃.	CaO.	MgO.	SO₃.	P₂O₅.	Sums.	Index to limestones.
		P. ct.	*P. ct.*	*P. ct.*	*P. ct.*	*P. ct.*	*P. ct.*	*P. ct.*	*P. ct.*	*P. ct.*	
4581	1	....	43.10	1.55	.97	54.70	.04	.03	.01	100.40	Selected from 150 cords of limestone on hand at the beginning of the season.
4582	2	.20	30.90	23.03	3.85	42.05	.07	.03	.03	100.16	Do.
4583	3	.05	41.84	3.10	1.40	53.55	.04	.02	.03	100.03	Do.
4584	4	......	37.82	3.00	1.48	51.3C	......	......	......	96.60	Core of limestone burned in mill, doesn't burn.
4585	5	....	40.07	5.92	.99	51.53	......	......	......	98.51	Limestone brought in wagons from Fort Scott Lime Works.
4586	6	....	37.56	10.01	1.07	49.26	......	......	......	97.90	Duplicate sample.
4598	7	....	26.06	2.50	1.56	63.84	.54	.37	......	94.87	Selected from 150 cords of limestone on hand at the beginning of the season.
4509	8	......	41.70	2.81	1.82	54.08	.27	......	......	100.68	Do.
4638	9	......	41.70	1.68	1.42	55.40	.25	......	......	....	Duplicate sample.
4651	10	......	42.70	3.80	.93	52.02	.54	......	......	99.99	Limestone in use October 17, surface rock from a point 2 miles south of factory.
4652	11	......	41.80	4.90	.72	52.26	.54	......	......	100.22	Limestone in use October 17 deeper in quarry.
4662	12	.10	40.00	5.40	3.14	57.83	.56	......	.05	101.03	Do.
4663	13	.04	41.20	3.47	2.02	53.72	.44	......	.10	100.99	Do.

	Burnt lime.	Slag.	Spent bone-black.
Serial number..........................	4,587	4,588	4,600
	Per cent.	*Per cent.*	*Per cent.*
Water	9.60	0.00	1.15
Carbonic acid CO₂......................	0.00	Traces...	1.13
Insol. and silica (SiO₂).................	4.36	39.30	.72
Fe₂O₃, Al₂O₃.............................	3.70	31.50	....
CaO......................................	81.40	38.60	....
MgO	.09	Traces.........	....
SO₃.......................................	Undetermined ..	Traces.........	....
P₂O₅......................................	Undetermined ..	....	*35.80
Organic matter...........................	Undetermined ..	....	17.24
Bases	....	....	43.84
Sum......................................	99.15	99.40	99.88

* Equivalent to Ca (PO₄)₂, 79.46.   † Contains N.

Mill juices before October 1.

No.	Extraction.	Sp. gr.	Solids.	Sucrose.	Glucose.	Date.	Index to mill juices.
	P. ct.		P. ct.	P. ct.	P. ct.		
1.....	55.20	1.0773	18.7	13.25	....	Aug. 30	Early amber cane from west field.
2.....	53.33	1.0660	16.3	11.46	1.88	Aug. 31	Early amber cane from east field.
3.....	57.14	1.0329	13.1	7.20	3.46	Aug. 31	Link's hybrid.
4....	58.82	1.0574	14.1	7.50	4.35	Aug. 31	Early orange.
5.....	......	1.0710	17.2	14.73	....	Sept. 3	Early amber cane, juice extracted by hand.
28....	60.60	1.0770	18.6	9.47	4.95	Sept. 15	Early amber cane from east field, cut two days.
31....	60.00	1.0788	19.0	7.04	7.80	Sept. 16	Early amber cane, cut three days.
37....	51.60	1.0794	19.2	4.02	8.42	Sept. 17	Orange cane from wagons.
44....	......	1.0688	16.7	10.83	2.49	Sept. 18	Cane from carrier.
53....	45.16	1.0832	20.0	13.54	2.07	Sept. 19	Do.
61....	47.15	1.0734	17.8	11.48	3.58	Sept. 20	Amber cane from carrier.
70....	68.68	1.0770	18.6	12.11	2.44	Sept. 21	Amber cane from carrier, cut yesterday.
71 ..	56.10	1.0750	18.2	11.82	2.73	Sept. 21	Orange cane from carrier.
84....	55.27	1.0818	19.5	11.02	4.20	Sept. 22	Amber cane from carrier, cut two days.
85....	58.62	1.0888	21.2	14.70	2.77	Sept. 23	Amber cane from carrier, cut one day.
87....	53.12	1.0848	20.3	3.60	11.36	Sept. 23	Amber cane from carrier, cut three days.
88....	61.77	1.0718	17.4	9.49	5.33	Sept. 23	Cane from carrier.
89 ..	58.44	1.0638	15.6	9.74	2.16	Sept. 23	Link's hybrid from field.
95....	46.43	1.0776	18.7	13.53	2.41	Sept. 24	Cane from carrier.
97....	56.56	1.0675	16.4	11.50	2.80	Sept. 24	Cane like preceding, except badly lodged.
102...	50.37	1.0578	14.2	8.20	2.86	Sept. 25	Cane from carrier, (from lodged lot).
103...	59.18	1.0678	16.5	10.17	3.47	Sept. 25	Orange cane, cut to-day.
106½..	53.00	1.0726	17.6	12.40	1.90	Sept. 28	Cane from carrier.
112...	51.51	1.0684	16.6	10.41	4.08	Sept. 29	Do.
119...	56.10	1.0764	17.8	12.39	3.76	Sept. 30	Do.
Aver.	55.79	1.0723	17.56	10.49	4.01		
Coefficient purity....			59.73				

Mill juices after September 30.

No.	Extraction.	Sp. gr.	Solids.	Sucrose.	Glucose.	Date.	Index to mill juices.
	P. ct.		P. ct.	P. ct.	P. ct.		
126	61.76	1.0634	15.5	8.37	4.95	Oct. 1	Cane from carrier, stripped.
131	......	1.0842	20.2	14.50	1.77	Oct. 2	Cane from carrier.
138	54.54	1.0866	20.7	14.37	2.16	Oct. 3	Cane brought in cars from Hammond.
147	51.72	1.0680	16.6	10.50	2.60	Oct. 4	Amber cane from carrier.
150	51.35	1.0740	17.9	12.39	1.92	Oct. 4	Orange cane from carrier.
159	51.35	1.0710	17.2	10.65	3.27	Oct. 5	Cane from carrier.
169	56.60	1.0818	19.7	13.20	2.37	Oct. 5	Cane, amber, on car from Hammond.
170	57.70	1.0778	18.8	9.95	4.88	Oct. 5	Same, orange.
176	52.94	1.0801	19.3	2.11	......	Oct. 7	Amber cane from Hammond.
177	53.85	1.0748	18.1	6.67	......	Oct. 7	Same, orange.
180	55.55	1.0698	17.0	10.69	3.11	Oct. 7	Cane from car, same as two preceding, but better averaged samples taken from center of car, while the first samples were taken from the outside, amber.
181	53.12	1.0828	19.9	12.46	3.03	Oct. 7	Same, orange.
199	60.10	1.0678	16.5	9.10	4.36	Oct. 8	Orange cane from carrier (juice very red).
207	50.63	1.0640	15.6	9.07	3.84	Oct. 8	Cane from carrier, p. m.
213	58.08	1.0758	18.3	4.55	9.62	Oct. 9	Cane from carrier, a. m. (old cane).
222	58.82	1.0596	14.6	8.57	2.20	Oct. 9	Link's hybrid from field.
223	61.54	1.0556	13.7	7.72	3.38	Oct. 9	Orange from field.
224	60.00	1.0506	12.5	7.22	4.09	Oct. 9	Amber from field.
231	59.37	1.0766	18.5	7.02	7.74	Oct. 10	Cane from carrier, cut several days.
232	62.07	1.0764	18.5	8.66	3.04	Oct. 10	First fresh wagon-load lot in to-day.
241	57.90	1.0676	16.5	10.29	2.13	Oct. 11	Link's hybrid cane from Professor Swenson's, still green and not hurt by frost.
242	......	1.0618	15.1	8.60	3.25	Oct. 11	Cane from carrier, freshly cut, a. m.
249	61.90	1.0632	15.5	8.86	2.98	Oct. 11	Cane from carrier, p. m.
258	62.16	1.0588	14.4	6.65	4.72	Oct. 12	Cane from carrier.
259	57.14	1.0718	17.4	8.54	5.04	Oct. 12	Cane on car from Hammond, orange.
260	58.83	1.0736	17.8	10.51	3.27	Oct. 12	Cane on car from Hammond, amber.
267	59.09	1.0592	14.5	8.83	3.10	Oct. 12	Cane, amber, lot from Hammond by Dr. Wiley and Professor Swenson.
268	61.54	1.0832	20.0	14.11	1.95	Oct. 12	Same, orange.
269	63.41	1.0672	16.4	9.53	4.56	Oct. 12	Same, orange, No. 2.
276	53.63	1.0956	21.5	5.71	11.41	Oct. 13	Cane for experimental run, orange, taken from same cars as yesterday's samples.
277	57.25	1.0846	20.3	12.05	4.19	Oct. 13	Same, amber.
279	50.46	1.0656	16.0	7.68	5.17	Oct. 13	Sample from other two cars from Hammond, orange.
280	62.07	1.0646	15.8	7.27	5.52	Oct. 13	Same, amber.
281	60.71	1.0654	16.0	10.09	2.23	Oct. 13	Orange cane from Professor Swenson's.
284	52.04	1.0618	15.1	4.15	7.84	Oct. 13	First mill juice from experimental run, taking sample every hour, orange.

Mill juices after September 30—Continued.

No.	Ex-trac-tion.	Sp. gr.	Solids.	Su-crose.	Glu-cose.	Date.	Index to mill juices.
	P. ct.		*P. ct.*	*P. ct.*	*P. ct.*		
285	62. 50	1. 0608	14. 9	4. 64	7. 25	Oct. 13	Same, amber, taken at same time as above.
287	....	1. 0621	15. 2	5. 83	6. 07	Oct. 13	Second sample, orange.
288	50. 00	1. 0626	15. 3	9. 51	2. 44	Oct. 13	Second sample, Link's hybrid, from Swenson's.
290	60. 00	1. 0678	16. 5	9. 77	3. 75	Oct. 13	Third sample, orange, from Hammond.
294	56. 52	1. 0684	16. 6	8. 51	4. 12	Oct. 14	Cane from carrier, amber.
295	62. 86	1. 0629	15. 4	7. 88	4. 38	Oct. 14	Cane from carrier, orange.
310	....	1. 0580	14. 3	7. 45	4. 50	Oct. 15	"Denton" cane, analyzed for Mr. Parkinson.
311	....	1. 0400	10. 0	3. 06	3. 31	Oct. 15	Green cane from wagon.
442	61. 58	1. 0560	13. 8	5. 87	4. 98	Oct. 23	Cane from field across railroad, amber, still gre
458	60. 00	1. 0550	13. 6	7. 60	1. 97	Oct. 25	Cane from field this side railroad track, amber.
Av.	58. 01	1. 068	16. 6	8. 70	4. 15		
Coefficient purity ..			52. 41				

Chips from first of season to October 1, 1886.

No.	Date.	Uncorrected.		Corrected.	
		Sucrose.	Glucose.	Sucrose.	Glucose.
		Per cent.	*Per cent.*	*Per cent.*	*Per cent.*
10	Sept. 8	5. 63	8. 34	5. 91	8. 02
11	Sept. 9	10. 21	3. 90	10. 72	3. 39
12	Sept. 9	9. 08	2. 15	9. 53	1. 70
14	Sept. 10	5. 86	5. 94	6. 15	5. 65
15	Sept. 11	9. 57	2. 17	10. 05	1. 69
17	Sept. 11	9. 90	1. 91	10. 40	1. 41
19	Sept. 13	9. 30	1. 24	9. 76	. 77
24	Sept. 14	8. 53	1. 94	8. 96	1. 51
36	Sept. 16	10. 32	3. 13	10. 84	2. 61
43	Sept. 17	9. 50	3. 34	9. 97	2. 86
49	Sept. 18	10. 81	2. 73	11. 35	2 19
65	Sept. 20	8. 28	4. 98	8. 69	4. 57
74	Sept. 21	7. 94	3. 11	8. 32	2. 73
86	Sept 22	5. 12	7. 14	5. 38	6. 88
90	Sept. 23	7. 39	4. 31	7. 75	3. 95
105	Sept. 25	5. 74	3. 89	6. 03	3. 60
107	Sept. 28	8. 51	2. 45	8. 93	2. 03
115	Sept. 29	11. 18	1. 97	11. 74	1. 41
121	Sept. 30	7. 26	6. 44	7. 62	6. 08
Mean		8. 43	3. 72	8. 85	3. 32

Chips from October 1 to close.

No.	Date.	Sucrose.	Glucose.	Sucrose.	Glucose.
129	Oct. 1	7. 88	4. 03	8. 28	3. 62
135	Oct. 2	8. 75	3. 57	9. 19	3. 13
151	Oct. 4	6. 89	3. 96	7. 23	3. 62
135	Oct. 7	8. 25	4. 22	8. 66	3. 81
206	Oct. 8	6. 22	6. 66	6. 53	6. 35
214	Oct. 9	6. 77	6. 66	7. 11	6. 32
227	Oct. 9	10. 73	3. 79	11. 27	3. 25
238	Oct. 10	7. 21	4. 88	7. 57	4. 52
246	Oct. 11	9. 08	3. 93	8. 48	3. 53
270	Oct. 12	9. 85	3. 26	10. 24	2. 77
286	Oct. 13	3. 91	7. 50	4. 11	7. 30
289	Oct. 13	5. 89	3. 83	6. 18	3. 54
297	Oct. 14	6. 65	4. 71	6. 98	4. 38
309	Oct. 15	7. 81	2. 87	8. 20	2. 48
315	Oct. 15	7. 31	3. 22	7. 68	2. 85
325	Oct. 16	7. 48	3. 61	7. 86	3. 23
341	Oct. 17	6. 55	4. 99	6. 88	4. 66
354	Oct. 18	5. 56	4. 14	5. 84	3. 86
375	Oct. 19	6. 16	4. 20	6. 47	3. 89
391	Oct. 20	6. 65	3. 93	6. 98	3. 60
413	Oct. 21	5. 77	4. 19	6. 06	3. 80
431	Oct. 22	4. 89	4. 55	5. 13	4. 31
447	Oct. 23	4. 62	4. 65	4. 85	4. 42
461	Oct. 26	4. 51	5. 47	4. 74	5. 24
474	Oct. 27	2. 48	5. 47	2. 69	5. 26
Mean		6. 08	4. 48	7. 01	4. 15

ANALYSES OF JUICE OF CHIPS FROM CUTTERS.

These chips were taken from the cells of the battery as they were filling. A handful was taken from each cell until ten had been sampled. The determinations were made by passing these chips through the mill and then subjecting the juice to examination in the usual way.

Mill juices from chips taken from circuit of cells.

Number.	Date.	Specific gravity.	Solids.	Sucrose.	Glucose.
			Per cent.	Per cent.	Per cent.
308	Oct. 15	1.0624	15.3	9.02	2.61
312	Oct. 15	1.0610	14.9	7.84	3.42
326	Oct. 16	1.0670	16.3	9.29	3.35
340	Oct. 17	1.0648	15.8	8.17	3.68
355	Oct. 18	1.0584	14.3	7.21	3.31
372	Oct. 19	1.0596	14.6	7.69	3.31
390	Oct. 20	1.0648	15.8	8.82	3.48
412	Oct. 21	1.0590	14.5	7.48	3.31
429	Oct. 22	1.0618	15.1	6.17	4.18
445	Oct. 23	1.0510	12.6	5.77	4.44
460	Oct. 26	1.0580	14.2	5.42	4.85
473	Oct. 27	1.0578	14.2	4.50	4.95
Means		1.0605	14.8	7.28	3.74
Means in cane..			13.17	6.48	3.31

Purity coefficient of juice, 49.
Glucose per 100 sucrose in juice, 51.07.

Chips exhausted in bottles with and without neutralizing.

Number.	Date.	Without addition of a neutralizing substance.		Expressed juice from same.		Neutralized by—	Gives—	
		Sucrose.	Glucose.	Sucrose.	Glucose.		Sucrose.	Glucose.
		Per cent.	Per cent.	Per cent.	Per cent.		Per cent.	Per cent.
354	Oct. 18	5.56	4.14	6.00	3.98	12 cc. ₁₀ alk.....	6.60	3.41
375	Oct. 19	6.16	4.20	6.60	4.10	16 cc. ₁₀ alk.....	7.31	3.23
391	Oct. 20	6.65	3.93	6.65	3.87	20 cc. ₁₀ alk.....	6.71	3.65
413	Oct. 21	5.77	4.19	6.10	4.48	Excess CaO....	6.55	.76
431	Oct. 22	4.89	4.55	5.61	4.44	1 cc. bisulphite..	4.84	4.80
447	Oct. 23	4.62	4.65	4.95	4.72	Excess CaCO₃ ..	5.17	4.00
461-2	Oct. 26	4.48	5.42			1 cc. CaCO₃ ...	5.11	5.31
461-2	Oct. 26	4.48	5.42			Excess CaCO₃ ..	4.68	5.03
461-2	Oct. 26	4.48	5.42			20 cc. ₁₀ alk.....	5.06	5.11
474-5	Oct. 27	2.73	5.63			2 grams CaCO₃..	3.33	4.95
Means		5.11	4.59	5.98	4.26		5.54	4.03

Number.	Expressed juice from same.		Mill juice from same chips.			Acidity in chips before neutralizing (malic acid).	Acidity in chips after neutralizing (malic acid).
	Sucrose.	Glucose.	Sucrose.	Glucose.	Solids.		
	Per cent.	Per cent.	Per cent.	Per cent.	Per cent.	Per cent.	Per cent.
354	7.53	3.07	7.21	3.31	14.3	.197	Trace.
375	7.43	3.29	7.69	3.31	14.6	.164	Trace.
391	6.65	Lost.	8.82	3.48	15.8	.197	None.
413	6.60	Lost.	7.48	3.31	14.5	.197	None.
431	4.84	4.81	6.17	4.18	15.1	.181	.246
447	5.39	3.93	5.77	4.14	12.6	.156	.049
461-2			5.42	4.85	14.2	.164	.066
461			5.42	4.85	14.2	.164	.049
461			5.42	4.85	14.2	.164	None.
474-5			4.50	4.95	14.2	.214	.082
Means	6.41	4.27	6.63	3.94	14.4	.180	
In chips			5.90	3.51	12.82		

Diffusion juices to October 1.

Number.	Date.	Solids.	Sucrose.	Glucose.
		Per cent.	Per cent.	Per cent.
13........	Sept. 9...	6.8	3.29	1.39
16........	11...	8.5	3.94	1.99
23........	13...	9.3	6.50	1.66
25........	14...	11.7	7.47	1.53
27........	14...	11.2	6.17	1.42
29........	15...	12.6	6.36	2.84
32........	16...	10.8	5.71	1.82
38........	17...	10.4	5.62	1.66
46........	18...	11.9	6.59	3.18
51........	18...	11.7	6.94	1.82
57........	19...	11.8	5.66	2.85
64........	20...	10.8	4.37	3.36
69........	20...	12.3	5.59	3.46
77........	21...	11.8	5.76	2.89
91........	23...	11.8	6.78	2.19
94........	24...	9.2	4.81	1.84
98........	24...	10.7	4.53	2.23
101........	25...	9.6	6.06	1.52
104........	25...	8.9	4.13	1.28
108........	28...	9.7	5.68	1.67
114........	29...	12.0	6.76	2.92
118........	29...	12.0	6.37	2.65
122........	30...	14.8	7.22	4.16
Average..	....	11.77	5.75	2.32

Purity, 48.93.

Diffusion juices October 1 to close.

Number.	Date.	Solids.	Sucrose.	Glucose.
		Per cent.	Per cent.	Per cent.
128........	Oct. 1	14.8	8.60	3.25
132........	Oct. 2	13.7	7.01	3.32
133........	Oct. 2	13.9	7.68	3.10
134........	Oct. 2	13.2	7.18	2.75
139........	Oct. 3	12.9	5.89	3.96
140........	Oct. 3	12.7	6.51	3.65
141........	Oct. 3	12.9	6.47	3.52
149........	Oct. 4	9.8	4.80	2.38
152........	Oct. 4	9.6	4.71	2.47
155........	Oct. 4	11.5	5.42	3.28
160........	Oct. 5	12.3	6.21	3.34
163........	Oct. 5	13.0	6.44	3.58
166........	Oct. 5	12.2	5.78	3.40
171........	Oct. 5	12.2	6.03	3.23
179........	Oct. 7	13.3	6.13	4.41
182........	Oct. 7	12.7	5.46	4.23
183........	Oct. 7	12.2	5.19	4.23
184........	Oct. 7	12.2	4.50	4.41
201........	Oct. 8	12.5	5.40	4.12
205........	Oct. 8	11.8	5.29	3.98
216........	Oct. 9	12.2	4.04	4.65
217........	Oct. 9	11.3	4.08	4.07
229........	Oct. 10	10.8	4.06	3.45
237........	Oct. 10	11.2	4.86	3.30
244........	Oct. 11	10.3	4.10	3.43
247........	Oct. 11	10.3	4.32	3.15
254........	Oct. 11	10.9	4.53	3.09
261........	Oct. 12	13.1	5.76	3.96
262........	Oct. 12	12.2	4.82	4.06
271........	Oct. 12	11.9	5.44	3.41
296........	Oct. 14	12.7	5 30	2.14
300........	Oct. 14	11.6	4.92	3.54
313........	Oct. 15	9.1	3.24	2.32
327........	Oct. 16	11.6	5.14	2.98
328........	Oct. 16	11.2	4.96	2.94
339........	Oct. 17	11.7	5.51	3.08
356........	Oct. 18	10.8	4.38	2.90
357........	Oct. 18	9.8	3.64	2.96
371........	Oct. 19	9.9	4.08	2.53
373........	Oct. 19	10.4	4.38	2.94
389........	Oct. 20	10.9	3.72	3.91
395........	Oct. 20	7.2	2.33	2.08
404........	Oct. 20	9.5	3.58	2.71

Diffusion juices October 1 to close—Continued.

Number.	Date.	Solids.	Sucrose.	Glucose.
		Per cent.	Per cent.	Per cent.
410........	Oct. 21	11.2	3.97	3.96
417........	Oct. 21	11.8	3.77	4.44
428........	Oct. 22	10.6	4.41	3.31
430........	Oct. 22	10.1	3.95	3.37
435........	Oct. 22	10.3	3.91	3.43
441........	Oct. 22	10.3	3.82	3.76
444........	Oct. 23	10.1	3.67	3.77
453........	Oct. 23	10.1	3.41	3.63
468........	Oct. 26	8.8	2.93	2.97
478........	Oct. 27	7.8	2.94	2.55
Average..	....	11.34	4.90	3.39

Filtered carbonatated juices before October 1.

Number.	Date.	Sucrose.	Glucose.	Solids.
		Per cent.	Per cent.	Per cent.
18........	Sept. 11	4.66	1.28	8.8
22........	Sept. 13	6.94	1.04	10.5
33........	Sept. 16	5.96	1.50	11.1
41........	Sept. 17	5.48	.94	11.4
47........	Sept. 18	6.53	2.39	10.9
67........	Sept. 20	4.78	1.82	10.7
78........	Sept. 21	4.55	1.24	9.9
Average..	....	5.56	1.46	10.47

Filtered carbonatated juices after September 30.

194........	Oct. 7	5.73	3.11	13.0
210........	Oct. 8	5.99	3.07	12.4
226........	Oct. 9	5.07	3.23	12.0
235........	Oct. 10	4.75	2.50	10.6
243........	Oct. 11	5.07	3.89	11.9
252........	Oct. 11	4.72	2.42	10.8
263........	Oct. 12	6.20	2.88	12.6
301........	Oct. 14	5.95	3.22	12.7
335........	Oct. 16	5.82	2.48	11.8
347........	Oct. 17	5.82	2.17	11.6
361........	Oct. 18	4.79	2.31	10.2
382........	Oct. 19	4.49	2.00	9.5
400........	Oct. 20	4.09	3.48	11.1
420........	Oct. 21	4.49	3.44	11.4
439........	Oct. 22	4.93	2.94	11.2
451........	Oct. 23	3.83	3.43	9.9
469........	Oct. 26	3.19	2.20	8.1
Average..	....	4.90	2.87	11.20

Sulphured juices before October 1.

Number.	Date.	Sucrose.	Glucose.	Solids.
		Per cent.	Per cent.	Per cent.
34........	Sept. 16	5.96	1.70	11.6
42........	Sept. 17	5.77	1.02	12.7
48........	Sept. 18	6.65	1.53	11.4
68........	Sept. 20	4.53	2.13	10.7
79........	Sept. 21	4.55	1.24	9.9
96........	Sept. 24	4.60	2.46	10.6
Average..	....	5.34	1.65	11.15

Sulphured juices after September 30.

195........	Oct. 9	6.73	3.11	13.2
211........	Oct. 8	5.89	3.17	12.6
225........	Oct. 9	5.09	3.12	12.2
236........	Oct. 10	4.78	2.03	10.6
253........	Oct. 11	4.78	2.54	11.0
264........	Oct. 12	6.20	2.07	12.7
302........	Oct. 14	5.87	3.50	13.2
336........	Oct. 16	5.89	2.57	12.5
348........	Oct. 17	6.18	2.44	12.4
362........	Oct. 18	5.12	2.35	10.8
383........	Oct. 19	4.58	2.14	9.7
401........	Oct. 20	4.12	3.54	11.2
421........	Oct. 21	4.54	3.53	11.7
440........	Oct. 22	4.89	3.04	11.4
452........	Oct. 23	3.90	3.48	10.8
470........	Oct. 26	3.23	2.31	7.8
Average..	....	5.11	2.90	11.5

Waste waters, before October 1.

Number.	Date.	Sucrose.	Glucose.
		Per cent.	Per cent.
20..............	Sept. 13	0.00	0.00
58..............	Sept. 19	.24	....
76..............	Sept. 21	.16	.10
99..............	Sept. 24	....	.25
111.............	Sept. 28	.67	....
117.............	Sept. 29	.31	.26
124.............	Sept. 30	....	.14
Average.....	....	.27	.15

Waste waters, after September 30.

136..............	Oct. 2	....	.17
142..............	Oct. 3	....	.43
145..............	Oct. 3	....	.15
153..............	Oct. 4	....	.91
156..............	Oct. 4	....	.43
161..............	Oct. 5	....	.30
164..............	Oct. 5	....	.35
167..............	Oct. 5	....	.48
172..............	Oct. 5	....	.21
186..............	Oct. 6	....	.14
187..............	Oct. 6	....	.08
188..............	Oct. 6	....	.08
189..............	Oct. 6	....	.11
202..............	Oct. 8	....	.21
208..............	Oct. 8	....	.19
218..............	Oct. 9	....	Trace.
219..............	Oct. 9	....	.15
230..............	Oct. 10	....	.11
248..............	Oct. 11	....	Trace.
250..............	Oct. 11	....	.10
265..............	Oct. 12	....	.20
266..............	Oct. 12	....	.10
273..............	Oct. 12	....	.10
299..............	Oct. 14	....	.10
303..............	Oct. 14	....	.11
316..............	Oct. 15	....	.62
330..............	Oct. 15	....	Trace.
331..............	Oct. 16	....	Trace.
343..............	Oct. 17	....	Trace.
358..............	Oct. 18	....	None.
381..............	Oct. 19	....	Trace.
405..............	Oct. 20	....	Trace.
406..............	Oct. 20	....	.20
438..............	Oct. 22	....	Trace.
457..............	Oct. 23	....	Trace.
Average (35)	....	....	.17

Waste chips, before October 1.

Number.	Date.	Glucose.
		Per cent.
26.................	Sept. 14	.15
35.................	Sept. 16	.20
40.................	Sept. 17	.10
52.................	Sept. 18	.10
56.................	Sept. 19	.10
81.................	Sept. 21	.23
92.................	Sept. 23	.20
106.................	Sept. 25	.38
110.................	Sept. 28	.29
116.................	Sept. 29	.63
Average....	....	.24

Waste chips, after September 30.

137.................	Oct. 2	.40
143.................	Oct. 3	1.20
144.................	Oct. 3	.71
154.................	Oct. 4	.79
157.................	Oct. 4	1.11
162.................	Oct. 5	1.52
165.................	Oct. 5	.70
168.................	Oct 5	.85
173.................	Oct. 5	.65
190.................	Oct. 7	.65
191.................	Oct. 7	.40
192.................	Oct. 7	.34
193.................	Oct. 7	.30
203.................	Oct. 8	.35
209.................	Oct. 8	22
220.................	Oct. 9	.55
221.................	Oct. 9	.41
230.................	Oct. 10	.42
251.................	Oct. 11	.02
256.................	Oct. 11	.22
272.................	Oct. 12	.62
274.................	Oct. 12	.34
275.................	Oct. 12	.30
298.................	Oct. 14	.41
304.................	Oct. 14	.39
314.................	Oct. 15	.56
329.................	Oct. 16	.18
345.................	Oct. 17	.24
359.................	Oct. 18	Trace.
379.................	Oct. 19	.69
396.................	Oct. 20	.30
418.................	Oct. 21	.42
436.................	Oct. 22	.36
454.................	Oct. 23	.41
Average (33).	....	.52

Semi-sirup.

Number.	Date.	Solids.	Sucrose.	Glucose.	Specific gravity.
		Per cent.	*Per cent.*	*Per cent.*	
62.................	Sept. 20	37.5	23.02	5.13	1.1660
204.................	Oct. 7	60.2	32.10	17.24	1.2910
228.................	Oct. 10	51.1	27.50	15.22	1.2388
255.................	Oct. 11	30.0	15.80	9.00	1.1296
291.................	Oct. 14	46.4	25.20	11.37	1.2130
307.................	Oct. 15	55.9	27.90	17.86	1.2663
319.................	Oct. 15	55.9	31.70	12.50	1.2600
337.................	Oct. 17	65.7	34.60	18.21	1.3248
353.................	Oct. 18	37.5	19.90	10.75	1.1664
366.................	Oct. 18	57.5	30.40	13.90	1.2750
425.................	Oct. 22	58.4	24.60	18.87	1.2800
456.................	Oct. 23	50.5	22.10	16.39	1.2356
Average (12).	....	50.5	26.23	11.94	

Masse-cuites.

Number.	Water.	Solids.	Ash.	Sucrose.		Reducing sugar.
				Direct.	Inversion.	
	Per cent.	*Per cent.*	*Per cent.*	*Per cent.*	*Per cent.*	*Per cent.*
12	18.98	81.02	04.35	46.20	47.84	22.72
40	19.41	80.59	05.09	40.00	41.14	21.51
45	18.99	81.91	04.6	43.60	45.82	24.04
57	19.34	80.66	08.4	44.40	46.92	24.75
92	17.60	82.40	04.9	44.20	46.93	21.93
306	16.69	83.31	04.94	42.60	44.98	21.93
332	35.14	64.86	05.46	47.10	48.73	19.53
338	17.77	82.83	04.45	45.70	46.97	20.83
350	22.21	77.79	05.03	38.80	40.92	20.32
370	14.58	85.42	05.43	47.50	48.20	21.19
387	16.94	83.06	03.16	48.80	50.42	16.56
Means	19.75	80.35	5.07	44.45	46.26	21.39

Molasses.

4683	26.42	73.58	.0539	32.80	34.48	21.33
4685				32.60		42.37
4681	18.07	81.93	.6540	35.10	36.94	20.83
4684	36.72	63.28	.0506	33.50	37.17	25.25
4686	23.12	76.88	.0502	30.3	33.41	28.10

Sample of sugar: Sucrose, 98.16; glucose, .07.

Acidity in juices.

[Calculated as malic acid.]

Mill juices.			Diffusion juices.		
No.	Date.	Per cent.	No.	Date.	Per cent.
179	Oct. 8	.280	184	Oct. 7	.321
180	Oct. 8	.255	201	Oct. 8	.174
213	Oct. 9	.201	216	Oct. 9	.263
232	Oct. 10	.280	229	Oct. 10	.273
242	Oct. 11	.188	244	Oct. 11	.295
258	Oct. 12	.147	327	Oct. 16	.335
326	Oct. 16	.188	356	Oct. 18	.308
355	Oct. 18	.188	371	Oct. 19	.147
372	Oct. 19	.126	417	Oct. 21	.161
412	Oct. 21	.134	430	Oct. 22	.362
429	Oct. 22	.134	453	Oct. 23	.348
442	Oct. 23	.147		Means	.272
	Means.	.189			

Acidity in chips.

No.	Date.	Per cent.	No.	Date.	Per cent.
105	Sept. 25	.234	286	Oct. 13	.181
107	Sept. 28	.222	289	Oct. 13	.164
115	Sept. 29	.226	325	Oct. 16	.164
121	Sept. 30	.246	341	Oct. 17	.173
129	Oct. 1	.214	354	Oct. 18	.197
135	Oct. 2	.197	375	Oct. 19	.164
151	Oct. 4	.181	391	Oct. 20	.197
185	Oct. 7	.164	413	Oct. 21	.197
206	Oct. 8	.181	431	Oct. 22	.181
214	Oct. 9	.185	447	Oct. 23	.156
227	Oct. 9	.104	461	Oct. 26	.164
238	Oct. 10	.230	474	Oct. 27	.214
246	Oct. 11	.148		Means..	.192
270	Oct. 12	.246			

Moisture in chips and bagasse.

Fresh chips.		Exhausted chips.		Fresh bagasse.		Exhausted bagasse.	
No.	Per cent. moisture.	No.	Per cent. moisture.	No.	Per cent. moisture.	No.	Per cent. moisture.
206	71.59	203	89.58	......	....	......	....
234	74.18	226	88.35	......	....	......	....
238	75.82	239	89.68	......	....	......	....
246	77.80	251	89.76	......	....	......	....
270	73.07	272	88.87	......	....	......	....
286	74.53	......	....	......	....	......	....
289	75.33	......	....	......	....	......	. ...
297	74.60	298	88.94	......	....	......	....
308	73.70	314	88.62	......	....	......	....
325	73.58	329	90.43	......	....	......	....
341	73.10	345	88.86	......	....	......	....
354	76.37	359	87.57	307	57.79	308	67.73
375	77.57	379	84.89	384	62.91	385	66.57
391	78.15	396	86.41	402	57.27	403	63.74
413	71.77	418	86.73	422	56.94	423	63.06
461	76.36	......	....	......	....	......	....
431	76.15	454	87.71	......	....	......	....
447	74.73	......	....	......	....	......	....
Means.	74.95		88.31		58.73		65.28

Table showing weight of twelve press-cakes.

No.	Pounds.	No.	Pounds.
1	26	7	24
2	24	8	24
3	24	9	24
4	24	10	23
5	25	11	24
6	25	12	24.5
		Mean.	24.3

Moisture in filter-press cake.

No.	Per cent. moisture.	No.	Per cent. moisture.
278	45.24	369	45.98
203	45.61	388	44.54
305	47.06	411	48.88
324	45.63	424	44.11
342	45.71	446	45.39
349	52.84		
		Mean.	46.45

Press-cakes.

[In dry substance.]

Serial No.	Organic matter.	Carbonic acid CO_7.	Insoluble and silica.	Iron and alumina.	Lime.	Magnesia.	Phosphoric acid $P_2 O_5$.	Sulphuric acid SO_3.	Manganese.	Total.
	Per ct.	*Per ct.*	*Per ct.*	*Per ct.*	*Per ct.*	*Per ct.*	*Per ct.*	*Per ct.*	*Per ct.*	*Per ct.*
4589										
4650	19.14	29.03	2.18	7.23	37.87	1.32	2.05	65	Traces.	99.47
4646	....	26.55	1.01	5.72	43.31	.71	.88	....		
4647	....	32.63	1.67	5.01	43.79	.59	.62	1.05	....	
4648	....	31.49	1.67	3.15	42.67	.70	.84	.82	....	
4649	.....	31.07	1.68	4.53	42.45	.32	.29	.92	....	
4660	23.28	29.90	.36	1.65	41.90	.60	1.20	.55	....	99.44
4687	17.80	31.78	.67	1.85	44.51	.76	.63	.90	....	99.90
4657	17.09	33.44	.07	3.96	43.72	.33	.09	.81	....	99.72
4658	15.27	33.50	1.45	4.07	45.36	.46	.39	.85	. ...	101.25
4659	17.00	32.35	.67	1.55	44.45	.69	.63	1.14	....	99.35
4675	11.88	31.95	.58	1.80	45.09	.56	.00	.98	....	99.95

[In the organic matter.]

Number.	Sucrose.	Glucose.	Nitrogen equivalent to albumen.	Water in original cake.
	Per ct.	Per ct.	Per ct.	Per ct.
4589......	.00	.74	2.66	39.07
4650......	.00	.83	....	....
4646......	1.90	2.83	4.65	45.24
4647......	2.20	3.77	2.15	45.61
4648......	2.43	3.08	....	47.06
4649......	2.41	5.02	5.70	45.63
4660......	.93	.83	1.08	52.84.
4687......	.07	Trace.	.90	45.98
4657......	3.12	3.00	.91	44.54
4658......	2.21	4.56	1.59	48.88
4659......	1.50	3.50	1.08	44.11
4675......	2.41	4.05	.33	....

DISCUSSION OF THE DATA.

It is evident from the foregoing analyses of limestones that, with few exceptions, the quality of stone used was exceedingly poor. The importance of good stone is at once evident, since bad stone is liable to "hang up" in the furnace, give a poor quality of lime for the defecation, and a weak gas for the carbonatations.

The quality of the gas employed during the season was fairly good. At first, by feeding too much coke with the limestone, large quantities of carbonic oxide were produced. The carbonic dioxide formed at the bottom of the furnace was reduced to CO by the white-hot coke above. After the laborers learned the proper manipulation of the kiln no further trouble was experienced from this cause. The carbonic oxide was always accompanied by a peculiarly unpleasant odor, and made the laborers about the carbonatation pans dizzy and ill. One of them fainted from the effects of the gas on the day it contained the largest quantity of carbonic oxide.

The percentage of CO_2 in the gas from time to time during the manufacturing season is shown by the following analyses:

No.	Date.	CO_2.	Remarks.
		Per cent.	
2	Sept. 28	11.00	
3	Sept. 29	13.00	
	Sept. 30	15.50	
4	Oct. 2	10.06	Morning.
5	Oct. 2	15.80	Noon.
6	Oct. 2	14.0	Night.
7	Oct. 3	21.0	
8	Oct. 4	22.4	Morning.
9	Oct. 4	20.6	Afternoon.
10	Oct. 5	22.0	
11	Oct. 6	23.0	
12	Oct. 9	22.5	
13	Oct. 10	22.5	
14	Oct. 11	23.0	
15	Oct. 19	15.0	

It is seen that when the men had learned the proper use of the furnace the percentage of CO_2 was kept pretty constantly above 20. The analysis No. 15 was made a day after the fires in the furnaces had been stopped. It showed that when internal combustion alone was practiced the percentage of CO_2 rapidly decreased. A gas containing from 20 to 25 per cent. CO_2 is well suited to carbonatation.

VOLUME OF GAS EMPLOYED.

The double-acting pump for supplying gas to the pans had the following capacity:

	Inches.
Diameter of cylinder	17.5
Length of stroke	21.25

The mean rate of motion for the pump was 40 per minute; hence the total quantity of gas delivered per minute was 236 cubic feet.

The volume of CO_2 furnished per minute is obtained by multiplying the above number by the mean percentage of CO_2 in the gas, viz, 236 $\times$ .20 = 47.2 cubic feet.

In metric terms 47.2 cubic feet are equal to 1,336 liters.

With gas of a good quality, say 25 per cent. CO_2, a pump of the capacity described would easily furnish gas for working 200 tons of cane per day.

DOUBLE CARBONATATION.

A few experiments were made to determine whether or not double carbonatations could be practiced with sorghum juices.

It was found that if from two to four tenths grams of lime per liter were left in the juice of the first carbonatation the filtration took place more readily, and the juice was somewhat purer.

In double carbonatation some additional lime is added to the hot juice from the filter-presses, and the injection of CO_2 continued until the liquid is neutral. Pans were put up and this method given a trial. But with a sugar-juice as rich in glucose as that afforded by sorghum, this procedure is not applicable.

For convenience, and to note the effects of a heavy frost, the analytical data relating to the juices, &c., are given in two parts, viz, those obtained before October 1 in the first part, and those after September 30 in the second. It is believed that every analysis made has been recorded, since in the circumstances arising from the result of the experiments even those which seem to have no value have been considered worthy of finding a place.

MILL-JUICES.

The samples of cane expressed by the small mill were taken without any purpose of illustrating any theory. The object in selecting them was to get as fair an idea as possible of the character of the cane entering the factory.

A study of the tables reveals the most surprising variations in the composition of the canes, varying from a quality of high sugar-producing value to one wortbless for this purpose.

As has already been pointed out, the generally poor character of the cane is due to much of it being overripe, especially in the case of the Amber variety. But the chief trouble arose from delay in handling the cane due to defects in the machinery already pointed out. In some cases, however, canes cut for two or three days, when kept, for example, in the middle of a car-load, from changes of temperature, preserved their sugar contents remarkably well. In general, however, the results of the work emphasized the importance of a prompt handling of the canes after they have been cut.

With such canes as are indicated by the analyses of the mill-juices it would be hopeless to expect to manufacture sugar profitably by any process whatever.

The amount of glucose per hundred of sucrose in the first series of analyses is 38.21; after September 30 it is 47.72.

DIRECT EXTRACTION OF THE CHIPS.

The determination of the sugars in the expressed juice of the cane is not a satisfactory method of determining the sugar in the cane itself. Did all canes contain the same percentage of juice, and were all the juice both that expressed and that remaining in the canes, of the same composition, no other method of analysis would be necessary. Since neither of these conditions obtain, however, in actual experience, I was led to try some other process. The one finally adopted is described in full in the Bulletin de l'Association des Chimistes, and published in Paris November 15, 1884.

Fresh sorghum-canes were cut into fine chips and treated for an hour in a closed bottle with water at the boiling temperature.

The analyses of the liquid obtained showed that the chips had the following composition:

No.	Sucrose in cane, by direct estimation.	Sucrose in cane, calculated from composition of the juice, 89 per cent.	Glucose in cane, direct.	Glucose in cane, calculated.
	Per cent.	Per cent.	Per cent.	Per cent.
1	*8.71	8.68	1.95	2.07
2	†7.98	7.82	1.84	1.86

* Mean of six analyses.
† Mean of four analyses.

It is seen by these analyses that the results obtained by the two methods agree very closely.

A large number of experiments has also shown that equally as satisfactory results are obtained with sugar-cane.

When, however, in the case of sorghum, the canes have already begun to deteriorate, and the sucrose is already partly inverted, it is found that this method of analysis causes a considerable inversion. A similar inversion, although to a less extent, takes place in the cells of the battery.

After the close of the season a comparative study was made of the amount of this inversion, and the results of these studies show clearly that the trouble is due to the acids of the cane chiefly to those formed by the partial fermentation which has produced the inversion of the sugar, or else the increased susceptibility of the sucrose remaining to the inverting action of the organic acids.

The results of these analyses are given under "analytical data."

DIRECT ESTIMATION OF SUGAR IN CHIPS.

The samples were taken as before described. Since only a small quantity could be used in each analysis (50 grams, circa), single results are not strictly mean indications of the content of the whole in sugar. The means, however, will give a fair idea of the composition of the chips.

The extraction of the sugar was made in the following way:

The weighed sample of fresh chips (48.9 grams) is placed in a strong extraction-flask and water added until the total volume (marked on neck of flask) is 305 cubic centimeters. The five cubic centimeters in excess of 300 is the allowance made for the fiber of the cane, which, for the quantity taken, amounts to five grams, and occupies a volume of about 5 cubic centimeters. The bottle is then tightly stoppered and heated at 100° for an hour, being frequently shaken. The method is based on the supposition that by this treatment complete diffusion has taken place, and that the free liquor and that in the pores of the pulp have the same composition. The liquor is then filtered, 100 cubic centimeters representing 16.3 grams of the original chips, treated with acetate of lead, made up to 110 cubic centimeters, and polarized. After adding one tenth the reading gives the percentage of sucrose present.

A discussion of the errors attending this method of analysis will be given further along.

Following are the numbers obtained by this method of analysis, and also the provisional correction which has been adopted.

This method rests on the assumption that the liquor within and without the chips has the same constitution. This assumption is probably incorrect when the canes have deteriorated.

Subjected to an analytical test the following data were obtained:

No.	Sucrose in liquor.	Sucrose in juice from chips.	Glucose in liquor.	Glucose in juice from chips.
	Per cent.	Per cent.	Per cent.	Per cent.
1	6.65	6.65	3.93	3.87
2	5.77	6.10	4.19	4.48
3	4.80	5.61	4.55	4.44
4	7.81	8.19	2.87	2.57
5	7.48	8.31	3.61	3.33
6	6.55	6.55	4.09	4.04
7	5.56	6.00	4.14	3.93
8	6.60	7.53	3.41	3.07
9	6.16	6.60	4.20	4.10
10	7.31	7.43	3.23	3.29
11	6.71	6.65	3.65	....
12	4.84	4.84	4.80	4.81
13	4.62	4.95	4.65	4.72
14	5.17	5.39	4.00	3.93
Means	5.75	6.06	3.69	3.87

Parts of glucose per 100 sucrose in free liquor, 69.94.
Parts of glucose per 100 sucrose in juice from the chips, 55.60.

It is seen from the above data that the mean total sugar in the free liquor equals 9.44 per cent. and in the juice expressed from chips from same equals 9.43 per cent.

This method of extraction with sorghum chips is, therefore, open to the objection of inverting a portion of the sucrose when the canes are not fresh. It is seen that 4 per cent. of sucrose present has been changed into reducing sugar. As the second of the analyses shows, this change has taken place entirely without the cell, the composition of the juice remaining in the cells being sensibly the same as that of the normal juice of the cane.

These results are of extreme interest. They show most conclusively that in the process of diffusion at a high temperature there is a notable inversion of the sucrose when the canes are not in proper condition. Further than this, it is shown that this inversion takes place in the sugar in the free liquor and not in the sugar remaining in the fiber of the cane. In nearly every case the free liquor was poorer in sucrose and richer in glucose than that in the pulp.

To correct the acidity in the battery, and thus avoid inversion, the following methods were tried:

(1) The limed juice used in the carbonatation-tanks was added to the cell of fresh chips little by little until enough was used to neutralize the acid. Two serious objections were found to this procedure: (a) The proper control of the quantity to be added was impossible. The juice would at times become strongly alkaline and highly colored; (b) the lime seemed to prevent the extraction of the sugar. The total solids of the diffusion juice under this treatment ran down rapidly from 11 per cent. to 4 per cent. This was due either to the coagulated albuminous matters preventing the osmotic action or to the formation of an insoluble lime sucrate, which remained in the chips. The method, therefore, had to be abandoned.

(2) Lime-water was added to the tank supplying the diffusion battery in such proportions as to furnish alkali enough to nearly neutralize the free acidity of each cell of chips. This water entered the cell next to be emptied of exhausted chips. All the lime in suspension was at once filtered out, and that in solution was not sufficient to neutralize the acidity in the cells in advance.

(3) Addition of lime bi-sulphite. To test the efficiency of lime bi-sulphite in preventing inversion during extraction it was added to the water in the feed-tank for the battery in quantities equal to one-half gallon for each diffusion. It was also used in the extraction flask with the results to follow.

(4) The addition of freshly precipitated carbonate of lime to the extraction bottle. This method was suggested by Prof. M. Swenson. The analyses show that the acidity was diminished by two-thirds, and the inversion of the sucrose largely prevented by the treatment. If a few pounds of such a carbonate could be evenly distributed in the

chips, it appears reasonable to suppose that this inversion would not take place.

Analytical data obtained in above experiments.

	Sucrose.	Glucose.	Exhausted chips, total sugars.
	Per cent.	Per cent.	Per cent.
Diffusion juice with lime in cells of fresh chips..	2.33	2.08	2.58
Compare analyses of same day of diffusion juice before the addition of lime	3.72	3.91	.57
Fresh chips used same day :			
Ordinary method.	6.65	3.93	....
Expressed juice from above	6.65	3.87	....
With alkaline extraction, NaO	6.71	3.65	....
Expressed juice from same	6.65	lost.	....
Mill juice from fresh chips same day	8.82	3.48	....

Per cent.
The diffusion juice from diffusion before treatment had of total sugars 7.63
Exhausted chips 51

Total sugar 8.14

The diffusion juice from chips treated with lime:
Total sugars 4.41
In exhausted chips 2.58

Total sugar 6.99

Whence it appears that by the coagulated albumen occluding the pores of the cells there was a loss of about 2 per cent. of sugar and in addition a small loss due to the formation of a lime sucrate.

In the extraction bottle, when the alkalinity was produced by lime instead of soda, this loss of sugar did not appear. The lime, however, diminished the percentage of glucose in a marked degree. This is shown by the following analyses:

	Sucrose.	Glucose.
	Per cent.	Per cent.
Extracted with water	5.77	4.19
Juice from the chips	6.10	4.48
Extracted with lime water	6.55	.76
Juice from the chips	6.60	not determined.
Diffusion juice made by adding lime to supply tank of battery :		
First	3.95	3.37
Second	4.41	3.31
Third. Diffusion juice with bi-sulphite added to supply tank, half gallon for each cell	3.91	3.43
Using bi-sulphite in extraction flask the following data were obtained :		
Mill juice from fresh chips	6.17	4.18
Ordinary extraction	4.80	4.55
Juice pressed from chips of above.	5.61	4.44
Extraction with addition of 1 cc. bi-sulphite to each bottle	4.84	4.86
Juice pressed from chips of above.	4.84	4.81

	First comparison.		Second comparison.	
	Sucrose.	Glucose.	Sucrose.	Glucose.
	Per cent.	Per cent.	Per cent.	Per cent.
Usual method	5.56	4.14	6.16	4.20
With alkali	6.60	3.41	7.31	3.23

In these two cases there was an apparent inversion of 20 per cent. of the sucrose. Another trial with better chips gave the following results:

	Sucrose.	Glucose.
	Per cent.	*Per cent.*
Chips treated in the usual way	6. 65	3. 93
Treated after addition of 20 cc., one-tenth alkali...........................	6. 71	3. 65
(In this case the inversion was only 1 per cent.)		
Another trial with very poor and sour chips:		
By direct method..................	4. 51	5. 47
With 1 cc CaCo₃.................`...`......	5. 11	5. 68
Sulphite	5. 11	4. 95
Means..............................	5. 11	5. 33
(Showing an apparent inversion of 10 per cent.)		
Same chips with an excess of CaCo₂	4. 68	5. 03
(Showing an apparent inversion of 3 per cent.)		

Taking all the data into consideration, it appears to be fair to assume that the inversion during the extraction in the flask was not more than 5 per cent. of the sucrose present, while during the first of the season it was doubtless much less. A strong corroboration of the justice of this allowance is found in the fact that the purity of the chips analyzed up to October 1, with the correction noted, is nearly exactly the same as that of the mill juices.

In the diffusion battery, where the temperature was kept at about 70° C., the inversion was not so great.

In any case, however, these analyses can only be accepted provisionally. The reliable analyses are those of the mill and diffusion juices. Since the results for the chips, however, agree so closely with those known to be correct, they can be accepted for all practical purposes.

Since the extraction in a flask does not afford a direct method of determining the total soluble solids in the chips, this must be done by calculation.

For this purpose the same ratio between glucose and other substances not sugar in solution is taken as that existing in the corresponding mill juices.

Applying this principle, we find that up to October 1 the following data are accessible.

                                                 Per cent.

Glucose in mill juices......`.`... 4. 01

Solids not sugar in mill juices.....`,`... 3. 06

Ratio, 1 glucose to .76.

Glucose in chips ....`.`... 3. 32

Not sugars calculated 2. 52

Sucrose.. . 8. 85

Total solids in chips 14. 69

After September 30 the numbers are as follows:

	Per cent.
Glucose in mill juices	4.15
Not sugars in mill juices	3.75

Ratio 1 glucose to .90 not sugars.

	Per cent.
Glucose in chips	4.15
Not sugars in chips (calculated)	3.74
Sucrose	7.01
Total solids	14.90
Purity of chips before October 1	60.5
Purity of chips after September 30	47.1

SAMPLES OF CHIPS—CORRECTED NUMBERS.

A full discussion of the data obtained by the analyses of the chips entering the battery has already been given.

Per hundred parts of sucrose the glucose was as follows:

	Per cent.
Before October 1	37.52
After September 30	59.18

A comparison of these ratios with those of the mill juices affords a confirmation of the supposition already expressed that as the canes deteriorate the rate of inversion on heating in a closed flask is greatly increased.

The analyses, therefore, of the mill juices after September 30 give the only fair idea of the character of the cane worked up to October 15. After that date the analyses of the juice of the chips pressed out by the experimental mill gives the best results possible. Sampled as the chips were, by taking an equal portion from each cell and mixing these sub-samples from ten cells together, the juice expressed therefrom is a fair representation of the character of the chips entering the battery.

JUICE FROM CHIPS PASSED THROUGH EXPERIMENTAL MILL.

From the analyses of the juices it is seen that the chips entering the battery from October 15 to the close of the season contained:

	Per cent.
Sucrose	6.48
Glucose	3.31
Glucose per hundred of sucrose	51.07

Leaving out of the computation the analyses of the chips in closed bottles, the following mean character of the cane for the entire season is obtained:

	Total solids.	Sucrose.	Glucose.
	Per cent.	Per cent.	Per cent.
Before October 1	15.63	9.34	3.57
After September 30	14.77	7.74	3.79
After October 14	13.17	6.48	3.31
Means	14.50	7.85	3.52

Mean purity, 53.9; mean glucose per hundred sucrose, 43.84.
Available sugar, calculated by taking difference between sucrose and all other solids, viz, 1.15 per, cent. = 23 pounds per ton.

It will be interesting to compare these numbers with those obtained at Magnolia Station, La., in 1885, and recorded in Bulletin No. 11, pp. 11, 12.

Per cent.

Total solids in cane 14. 22
Total sucrose in cane 10. 90
Total glucose in cane .... .92
Mean purity 76. 6
Mean glucose per 100 sucrose 8. 44

·Available sugar calculated as before, viz, 7.58 per cent. =151.6 pounds per ton.

It thus clearly appears from a careful study of the analytical data that the sorghum canes entering the battery at Fort Scott were totally unfit for sugar-making. Those who are disposed to find fault with the experiments because more sugar was not made would do well to consider these facts.

No known process, save an act of. creation, could have made sugar successfully out of such material.

If nothing better than this can be obtained, then it is time to declare the belief in an indigenous sorghum-sugar industry a delusion. This subject will be mentioned again in the summary.

A general review of the data connected with this interesting problem shows that with fresh chips of fine quality, the natural acidity is capable of producing no appreciable inversion during treatment in an extraction flask or while under pressure in the battery. With the deterioration of the cane, however, and consequent increasing acidity, this inversion becomes very great. In other words, the natural acids of the cane, such as malic and aconitic, are incapable of producing any appre ciable inversion; but the accidental acid (acetic) which comes from deterioration may cause an inversion of the sucrose in a most marked degree. The most practical method of avoiding this danger appears to me to be a mechanical contrivance which will sprinkle evenly over the entering chips 2 or 3 pounds of fine slaked lime or double that quantity of fine calcium carbonate to each cell of chips.

As has already been noted, every other attempt to neutralize the dangerous acids of the cane in a practical way has failed.

DIFFUSION JUICES.

The ratio of glucose to sucrose (per hundred) in the diffusion juices was as follows:

Per cent.

Before October 1 .... 39. 95
After September 30 .... 68. 15

These results show that before frost the inversion of the sucrose in the battery was nil, but that after frost this inversion was very marked. This fact is also emphasized by another, viz, that before frost the full ·battery of 14 cells was used, but that afterwards 8, 10, and 12 cells only were employed. Thus before frost the chips in the battery were longer under pressure than afterwards, and I may add that the temperature was also higher. These facts corroborate the statement already made that when once the process of inversion has commenced it goes easily and rapidly forward under the combined influence of time and an elevated temperature. Before such deterioration begins a temperature of even 100° C, can be maintained for an hour without notable injury.

A further fact which is illustrated by the analyses of the diffusion juices from uninjured canes is that the diminished purity is produced solely by the extraction of gum and chlorophyll, chiefly from the blades and sheaths, and that this injury can be avoided by a proper cleaning of the canes.

With clean canes and those in which the sucrose is still uninjured no alkaline substance will have to be used in the battery. When, however, deteriorated canes are used, some such application will be necessary to save the sucrose from further inversion. As has already been pointed out, finely powdered lime or calcium carbonate evenly distributed over the chips offers the simplest solution of the difficulty.

CARBONATATED JUICES.

The ratio of glucose to sucrose (per hundred) was as follows:

	Per cent.
Before October 1	26.28
After September 30	57.40

In both cases we find a marked decrease in the quantity of glucose. This produces a corresponding increase, usually reckoned at twice the quantity of glucose destroyed, in the *rendement* of crystallized sugar.

If the resulting molasses could be preserved—and this can be done, as will be pointed out later—this increase in yield could be used without any deleterious effect whatever. The analytical data confirm the opinion already expressed, and agree with the experience of sugar-makers wherever the process has been tried, that the process of carbonatation gives a larger yield of crystallizable sugar than can be obtained by any other known method of defecation.

SULPHURED JUICES.

Comparing again the glucose per hundred of sucrose, the following data are obtained:

	Per cent.
Before October 1	30.86
After September 30	36.84

In the first part of the season the treatment with sulphurous acid shows a very slight inversion of the sucrose. This was accomplished by long treatment of the juice with the acid, in the hope that a lighter-colored sirup might be produced.

In the second half of the season no inversion took place from this source. As I will point out further along, the treatment of the juice at this point by sulphur should be replaced by the addition of phosphoric acid.

The sulphurous acid should be applied afterwards, but in the double effect and strike pans.

WASTE WATERS AND EXHAUSTED CHIPS.

The amount of waste water was very small, compressed air having been uniformily used to drive the water from the cell next to be discharged.

In the estimation of the sugar the sucrose was first inverted and the whole sugar estimated as glucose. The mean percentage of both sugars in the waste waters after September 30 was .17 per cent. Since the mean glucose per hundred of sucrose for the season was nearly 44, the respective quantities of sucrose and glucose were as follows:

	Per cent.
Sucrose	.11
Glucose	.06

In the exhausted chips before October 1, by the same method of calculation, there was of—

	Per cent.
Sucrose	.16
Glucose	.08

After September 30 the numbers are as follows:

	Per cent.
Sucrose	.35
Glucose	.17

This increase in the sugar left in the chips was due to cutting out a large portion of the battery, especially during the first week in October. At this time often only six cells were under pressure, but the result is seen in the large quantities of total sugar left in the chips, amounting in one instance to 1.52 per cent.

After the 6th of October nine or ten cells were kept under pressure, and the content of sugar left in the chips was correspondingly diminished.

Sorghum, however, lends itself to diffusion more readily than any other sugar-producing plant, and a battery of ten cells properly managed would give good results as far as extraction is concerned.

PRESS CAKES.

The mean weight of the press cakes was 24.3 pounds. The mean content of moisture was 46.45 per cent.

Since considerable time elapsed from the time of sending the cakes from Fort Scott until they were analyzed at Washington, a considerable inversion of the sucrose took place.

The mean total sugar in the twelve press-cakes examined was 4.42 per cent.

Dividing this, as before, between the two sugars, we find, of—

	Per cent.
Sucrose	2.97
Glucose	1.45

When extra care was taken in washing the cakes, as in the case of the Louisiana experiments, to be later described, only a trace of sugar was left in them.

A glance at the composition of the cake will show its value as a fertilizer.

The quantity of lime used was nearly 1½ per cent. of the weight of the cane entering the battery.

RESULTS OF WORK.

The average weight of chips in the cells was 1,900 pounds.

From the beginning of the first attempts to run the machinery (September 13) until it was found possible to save the product (September 29) 499 diffusions were made, amounting to 948,100 pounds. After beginning to save the product (September 29) until suspension of work (October 26) 1,945 diffusions were made, amounting to 3,695,500 pounds. The total weight of cane, seed, and blades received from the field after September 19 was 3,120 tons.

The weight of chips diffused was 2,322 tons. The weight of seed, tops, blades, and cleanings (by difference) was 798 tons.

Following is the number of cells of chips used each day after September 19. Before that date no separate daily account was kept:

Date.	Number of cells cut.	Date.	Number of cells cut.	Date.	Number of cells cut.
Sept. 20	30	Oct. 2	69	Oct. 14	80
21	59	3	56	15	75
22	44	4	79	16	100
23	67	5	55	17	85
24	89	6	53	18	55
25	63	7	66	19	53
26	66	8	59	20	91
27	41	9	70	21	102
28	33	10	79	22	106
29	75	11	92	23	99
30	60	12	85	26	42
Oct. 1	67	13	66		
Total					2,419

About one-third of the cane received was partly stripped of its blades. It appears from the above figures that the seed tops, blades, and sheaths of the cane will amount to nearly 30 per cent. of the entire weight. It must also be remembered that much of the blades, sheaths, &c., was not removed by the very imperfect cleaning apparatus employed, and this weight is included in that of the " clean chips."

Weight of cane taken...pounds.. 118,480
Weight of seed tops ....do... 21,875
Weight of cleanings...do.... 7,580
Weight clean cane chips...do.... 89,025
Weight of each cell full of clean chips...do.... 1,894
Seed heads to total weight of cane...per cent.. 18.47
Cleanings total weight of cane...do... 6.40
Clean chips on total weight of cane...do.... 75.13

The cane used in the above experiments was stripped in the field. The "cleanings" comprised the blades not removed and sheaths, &c., blown out by the fanning-machine. Much of these impurities was not removed. The sugar obtained was of a fair marketable kind and found a ready sale. The molasses was of a dark color and a poor quality.

The weight of *masse-cuite* was determined on a portion of the product by Mr. Swenson. He placed it at a mean of 12 per cent. of the weight of the chips entering the battery. The weight of melada obtained from the 2,322 tons was, therefore, 557,280 pounds, or 46,440 gallons.

At the present writing (November 15) all of the sugar has not been swung out, but the product will be about fifty thousand pounds. This is indeed a discouraging yield and quite in contrast with the phenomenal quantity obtained from sugar-cane from Louisiana, to be mentioned further along. If a proper crystallizing room had been provided by the company the yield of sugar would have been much larger. On November 2 the different parts of the crystallizing room were found to be of the following temperatures :

	Degrees F.
Northeast corner	84
North center	84
Three feet above floor, under north steam-drum	72
Northwest corner	75
In upper layer of sirup in wagon, under south steam drum	105.8
Bottom of same wagon	77
South center	79
Southwest corner, over office	79
Between steam-drums	80.1
Temperature of air outside in shade	64.4

At such a low temperature a *masse-cuite* poor in sucrose and boiled to string proof cannot crystallize to advantage.

Before beginning the experiments with sugar-cane about to be described I obtained permission of the company to provide a special hot room. With such material and with such unfavorable conditions of crystallization the yield of over 20 pounds of sugar per ton is a convincing proof of the efficiency of the process employed.

DISPOSITION OF THE EXHAUSTED CHIPS.

The problem of the disposition of the exhausted chips is one of great importance. By the failure of the machinery which was designed to remove the chips to a considerable distance from the building, the chips had to be taken away by scrapers. When it is remembered that these chips have slightly increased in weight in passing through the battery the great expense of this proceeding is at once apparent.

The percentage of water in the discharged chips was found to be as follows:

Number.	Per cent.		Number.	Per cent.
1	84.80		6	90.43
2	86.73		7	89.08
3	87.54		8	88.87
4	86.41		9	88.94
5	88.62		10	88.86
			Mean	88.007

Since the mean of former experiments shows that sorghum contains about 11 per cent fiber and matters insoluble in water, the composition of the waste chips as indicated by the above determination is:

	Per cent.
Fiber	11.00
Water	88.10
Other substances	.90
Total	100.00

After passing the waste chips through the mill they had the following per cent. of water:

Number.	Per cent.
1	60.57
2	83.74
3	63.06
4	67.73
Mean	65.28

At 70 per cent. extraction the bagasse therefore contains one part of fiber to two of water. By a short preliminary drying this bagasse would readily burn. At any rate it presses so readily, requiring so little power, that in my opinion, it would be a matter of economy to pass it through a three-roll mill.

The percentage of extraction obtained with the spent chips in small experimental mill will be seen by the following numbers:

The first column represents the per cents. calculated from weighing the bagasse and the second from weighing the expressed water:

Number.	From bagasse.	From water.
	Per cent.	Per cent.
1	73.00	70.65
2	72.10	68.31
3	80.00	64.35
4	72.80	69.20
5	70.80	65.33
Mean	73.76	69.17

Since it is difficult to accurately collect and weigh the fine bagasse which the spent chips afford, the mean of the second column will be found to represent more accurately the real extraction. It is certain that with a good three-roll mill each 100 pounds of the spent chips can be reduced to 30 pounds, one-third of which is combustible material. Even if no attempt is made to use the bagasse as a fuel the pressure is to be recommended on the score of economy. There appears to be no difficulty whatever in passing the chips through a three-roll mill, and their soft and pulpy state renders the pressure exceedingly easy.

Further reference to this point will be made in that part of the report devoted to sugar-cane.

THE CHARACTER OF THE CANE USED SEPTEMBER 27 TO OCTOBER 6, INCLUSIVE.

A considerable amount of interest has been excited by comparisons made of the cane worked during the time above mentioned and that used subsequently.

MILL JUICES.

The mill juices analyzed during this time had the following composition:

No.	Date.	Extraction.	Specific gravity.	Solids.	Sucrose.	Glucose.	Remarks.
		Per cent.		Per cent.	Per cent.	Per cent.	
106½	Sept. 28	53.00	1.0726	17.6	12.40	1.90	Cane from carrier.
112	Sept. 29	51.51	1.0084	16.6	10.41	4.08	Do.
119	Sept. 30	56.10	1.0764	17.8	12.39	3.76	Do.
126	Oct. 1	61.76	1.0634	15.5	8.37	4.05	Do. (stripped).
131	Oct. 2	....	1.0842	20.2	14.50	1.77	Cane from carrier.
138	Oct. 3	54.54	1.0866	20.7	14.37	2.10	Cane brought in cars from Hammond.
147	Oct. 4	51.72	1.0680	16.6	10.50	2.60	Cane, amber, from carrier.
150	Oct. 4	51.35	1.0740	17.9	12.39	1.92	Cane, orange, from carrier.
159	Oct. 5	51.35	1.0710	17.2	10.65	3.27	Cane from carrier.
109	Oct. 5	56.00	1.0818	19.7	13.20	2.37	Cane, amber, on cars from Hammond.
170	Oct. 5	57.70	1.0778	18.8	9.95	4.88	Cane, orange, on cars from Hammond.
	Mean...	54.50	....	18.1	11.74	3.06	

No analyses were made on September 27 nor October 6.

Per cent.
Mean purity of juice during time mentioned 64.8
*Mean purity of juice after October 6 .... 49.7
Mean glucose per hundred sucrose during time mentioned 26.67
Mean glucose per hundred sucrose after October 6 .... 54.68

* Total solids, 16.2 per cent.; sucrose, 8.05 per cent.; glucose, 4.41 per cent.

Diffusion juices.

Number.	Date.	Solids.	Sucrose.	Glucose.
		Per cent.	*Per cent.*	*Per cent.*
108	Sept. 28	0.79	5.68	1.67
114	Sept. 29	12.6	6.76	2.92
118	Sept. 29	12.0	6.37	2.65
123	Sept. 30	14.6	7.22	4.16
128	Oct. 1	14.8	8.60	3.25
132	Oct. 2	13.7	7.01	3.32
133	Oct. 2	13.0	7.68	3.10
134	Oct. 2	13.2	7.18	2.75
130	Oct. 3	12.0	5.89	3.96
140	Oct. 3	12.7	6.51	3.65
141	Oct. 3	12.0	6.47	3.52
149	Oct. 4	9.8	4.80	2.38
152	Oct. 4	9.6	4.71	2.47
155	Oct. 4	11.5	5.42	3.28
160	Oct. 5	12.3	6.21	3.34
163	Oct. 5	13.0	6.44	3.58
166	Oct. 5	12.2	5.78	3.40
171	Oct. 5	12.2	6.03	3.23
Mean		12.4	6.04	3.15

Mean purity of juice during time mentioned 48.7
*Mean purity of juice after October 7 .... 40.0
Mean glucose per hundred sucrose during time mentioned 52.13
Mean glucose per hundred sucrose after October 7 .... 77.77

The mean purity of the mill juices during the interval named was 64.8 and of the diffusion juices 48.7, a loss of 16.1 points.

During the rest of the season the mean purity of the mill juices was 49.7 and of the diffusion juices 40.0, a loss of only 9.7 points.

The glucose per hundred of sucrose, during interval noted, in the mill juices was 26.07. In the diffusion juices it was 52.13, an increase of 26.06 points. During the rest of the season the glucose per hundred of sucrose in the mill juices was 54.68; in the diffusion juices 77.77; an increase of 23.09 points.

The most striking point about these comparisons during the interval named is the enormous difference between the mill juices and those of diffusion. In no other part of the season does the deterioration of the juice in the battery show itself to such an alarming extent.

There is only one explanation of this which appears satisfactory, and that is the fact that during this time the temperature of all the cells under pressure except the two central ones was kept within the limits of fermentation. The cane during this period, as a glance at the analyses will show, was by far the best worked during the entire season. The analyses of the chips made during this time shows the following mean results:

	Uncorrected.	Corrected.
	Per cent.	*Per cent.*
Sucrose	8.41	8.82
Glucose	3.73	3.32

Corrected glucose per hundred of sucrose, 37.65.

* Total solids, 10.9 per cent. ; sucrose, 4.36 per cent. ; glucose, 3.44 per cent.

Thus, compared directly with the chips, the inversion in the battery was great.

Judged by the same standards, there was at no other time during the season so great an inversion of sucrose in the battery as during this period of few cells and low temperatures. Nevertheless the character of the cane was so good that the yield of sugar was large. Had, however, the cane been worked without the inversion spoken of, the yield of sugar would have been twice as large. During the same period the percentage of total sugars left in the exhausted chips was .80, while before this time it had only been .17.

It is therefore seen from the data given that the attempt to work the battery with few cells and at a low temperature increased the sugar left in the chips more than one-half, and caused a greater inversion of the sucrose than was experienced at any other time during the entire season.

I call especial attention to these facts, because during the period mentioned I was absent from Fort Scott. On my return I ordered the battery to be worked with nine or ten cells under pressure and at a uniform temperature of 70° C. This I believe to be the best method of operating a diffusion battery for sorghum, at least until some method is invented of distributing over the chips some substance which will neutralize the acids of the cane and thus entirely prevent inversion. The methods by which I attempted to accomplish this desirable result have already been described.

A further fact, which is illustrated by the analyses of the diffusion juices from uninjured canes, is that the diminished purity is produced solely by the extraction of gum and chlorophyll chiefly from the blades and sheaths, and that this injury can be avoided by a proper cleaning of the canes.

With clean canes and those in which the sucrose is still uninjured no alkaline substance will have to be used in the battery. When, however, deteriorated canes are used some such application will be necessary to save the sucrose from further inversion. As has already been pointed out, finely powdered lime or calcium carbonate evenly distributed over the chips offer the simplest solution of the difficulty.

MODIFICATION OF THE PROCESS OF CARBONATATION.

In order to avoid the discoloration of the sirup, which is the chief objection to carbonatation, the following modification of the process was adopted:

The juice used was obtained from sugar-cane sent from Fort Scott to Washington, and the experiments were made after my return from Kansas.

To the cane-juice was added 1 per cent. of its weight of freshly burned lime, and the carbonatation was continued until the juice was almost neutral. After raising to the boiling point to decompose sucro-carbonates the juice was filtered, and then enough phosphoric acid added to precipitate the lime remaining in solution.

Since a slight excess of the acid will redissolve the precipitate and form acid phosphate, sodium phosphate was substituted for the phosphoric acid.

Much of the red color of the carbonatated juice was discharged by this process. After the precipitation was complete the juice was again boiled and filtered. It was then bleached with sulphurous acid and evaporated to 40° B.

In every instance the sirup made in this way was very light in color, perfectly transparent, and of the finest flavor. So pure was it, indeed, that it was found unnecessary to use any acetate of lead or any other defecating material to prepare this sirup for polarization. The quantity of phosphate of soda required to precipitate the lime in 5 liters of juice (11 pounds) was 100 cubic centimeters of a 10 per cent. solution. Therefore 10 grams of the sodium phosphate are sufficient for 5,000 grams of juice. About 4 pounds of sodium phosphate or 3 pounds of phosphoric acid would be sufficient for working a ton of cane.

The whole cost of treating cane juices with phosphoric acid or sodium phosphate will not be over 15 cents per ton of cane. The phosphoric acid, however, is not lost. It will reappear in the press cakes, having lost only half its value. Hence the actual cost of using this method of removing the lime is not probably over half of the estimate given above.

I made every effort to get phosphoric acid at Fort Scott, but could not succeed in time.

I believe the modification of the process here suggested will make a noted improvement in the molasses over any other procedure now in use.

GENERAL CONCLUSIONS.

In a general review of the work, the most important point suggested is the absolute failure of the experiments to demonstrate the commercial practicability of manufacturing sorghum sugar. The causes of this failure have been pointed out in the preceding pages, and it will only be necessary here to recapitulate them. They were:

(1) Defective machinery for cutting the canes and for elevating and cleaning the chips and for removing the exhausted chips.

(2) The deterioration of the cane due to much of it becoming over-ripe, but chiefly to the fact that much time would generally elapse after the canes were cut before they reached the diffusion battery. The heavy frost which came the 1st of October also injured the cane somewhat, but not until ten days or two weeks after it occurred.

(3) The deteriorated cane caused a considerable inversion of the sucrose in the battery, an inversion which was increased by the delay in furnishing chips, thus causing the chips in the battery to remain exposed under pressure for a much longer time than was necessary. The mean time required for diffusing one cell was twenty-one minutes, three times as long as it should have been.

(4) The process of carbonatation, as employed, secured a maximum yield of sugar, but failed to make a molasses which was marketable. This trouble arose from the small quantity of lime remaining in the filtered juices, causing a blackening of the sirup on concentration, and the failure of the cleaning apparatus to properly prepare the chips for diffusion.

A modification of the process which will prevent this trouble has already been explained; but, although an earnest attempt was made to introduce this method, it was found impossible to accomplish it before the end of the season.

I doubt whether any other industry has ever been the object of so much misrepresentation as this one.

In the preceding report I have endeavored to lay before you all the facts noted in the recent experiments. If I have not interpreted them correctly, I have, at least, given the data for a correct interpretation.

I should, indeed, be glad to leave this industry in a more promising condition. All admit that the process of diffusion has been successfully worked out, and to this opinion I subscribe, with the reservation that a proper mechanical method for distributing over the chips a substance to prevent inversion of the sucrose has not yet been discovered.

Honest differences of opinion still exist in respect of the best method of treating the diffusion juices, but it has been shown at Rio Grande that the diffusion juice from clean cane can be worked without any purification whatever.

Whether this purification is to be accomplished by carbonatation, filtering with brown coal, or in some other way, can easily be decided without menacing the future of the sorghum industry.

The problem of successfully cutting and cleaning the canes does not appear to me to be incapable of solution. It should have been solved the first thing, without leaving it for the last.

Last of all, the chief thing to be accomplished is the production of a sorghum plant containing a reasonably constant percentage of crystallizable sugar.

I cannot emphasize this point better than by quoting from some of my previous reports. In Bulletin No. 3, pp. 107–108, the following words are found:

IMPROVEMENT BY SEED SELECTION.

I am fully convinced that the Government should undertake the experiments which have in view the increase of the ratio of sucrose to the other substances in the juice. These experiments, to be valuable, must continue under proper scientific direction for a number of years. The cost will be so great that a private citizen will hardly be willing to undertake the expense.

The history of the improvement in the sugar-beet should be sufficient to encourage all similar efforts with sorghum.

The original forage beet, from which the sugar-beet has been developed, contained only 5 or 6 per cent. of sucrose. The sugar-beet will now average 10 per cent. of sucrose. It seems to me that a few years of careful selection may secure a similar improvement in sorghum.

It would be a long step toward the solution of the problem to secure a sorghum that would average, field with field, 12 per cent. sucrose and only 2 per cent. of other sugars, and with such cane the great difficulty would be to make sirup and not sugar. Those varieties and individuals of each variety of cane which show the best analytical results should be carefully selected for seed, and this selection continued until accidental variations become hereditary qualities in harmony with the well-known principles of descent.

If these experiments in selection could be made in different parts of the country, and especially by the various agricultural stations and colleges, they would have additional value and force. In a country whose soil and climate are as diversified as in this, results obtained in one locality are not always reliable for another.

If some unity of action could in this way be established among those engaged in agricultural research, much time and labor would be saved and more valuable results be obtained.

In Bulletin No. 5, pp. 185–6–7, are found the following conclusions :

A careful study of the foregoing data will not fail to convince every candid investigator that the manufacture of sugar from sorghum has not yet proved financially successful.

The men who have put their money in these enterprises seem likely to lose it, and intending investors will carefully consider the facts herein set forth before making final arrangements. The expectations of the earlier advocates of the industry have not been met, and the predictions of enthusiastic prophets have not been verified. It would be unwise and unjust to conceal the facts that the future of the sorghum-sugar industry is somewhat doubtful. The unsatisfactory condition is due to many causes. In the first place, the difficulties inherent in the plant itself have been constantly undervalued. The success of the industry has been based on the belief of the production of sorghum with high percentages of sucrose and small amount of reducing sugar and other impurities.

But the universal experience of practical manufacturers shows that the average constitution of the sorghum-cane is far inferior to that just indicated. Taking the mean of several seasons as a sure basis of computation, it can now be said that the juices of sorghum as they come from the mill do not contain over 10 per cent. of sucrose, while the percentage of other solids in solution is at least 4.

It is needless to say to a practical sugar-maker that the working of such a juice is one of extreme difficulty, and the output of sugar necessarily small.

The working of sorghum juices will be found as difficult as those of beets, and true success cannot be hoped for until the processes used for the one are as complete and scientific as for the other. It is not meant by this that the processes and machinery are to be identical.

The chemical as well as mechanical treatment of the two kinds of juice will doubtless differ in many respects. And this leads to the consideration of the third difficulty, viz, the chemical treatment of sorghum juice. It has taken nearly three-quarters of a century to develop the chemistry of the beet-sugar process, and even now the progress in this direction is great. The chemistry of the sorghum-sugar process is scarcely yet a science. It is only an imitation of what has been done in other fields of work. Sorghum will have to develop a chemistry of its own. This will not be the work of a day or a year, but it will be accomplished sooner or later.

Careful study of climate and soil, joined with experience, will gradually locate those areas most favorable to the growth of this plant and its manufacture.

This is an all-important point in the problem, and is now occupying seriously the attention of the thoughtful advocates of the sorghum-sugar industry. One thing is already clear, i. e., that the area of successful sorghum culture is not nearly so extensive as it was thought to be a few years ago. I would urge a further investigation in this direction as a work peculiarly within the province of the Department, and one which would prove of immense benefit to the country. Five million acres of land, suitable to the purpose, will produce all the sugar required for this country for several years to come. It is therefore certain that the sugar industry will be confined to the most favorable localities. If a thorough, scientific study of all the soil and climatic conditions does not point out this region, bitter experience and the loss of hundreds of millions of dollars will gradually fix its boundaries. Last of all, the sorghum industry has suffered from the general depression which has been felt by the sugar industry of the entire world. Low prices have caused loss where every other condition has been favorable. It is hardly probable that the price of sugar will rise again to its maximum of the years passed. Only war, pestilence, or disaster would produce this effect. It is best, therefore, for the sugar-grower to accept the present price as final and make his arrangements accordingly. But low prices will produce increased consumption, and thus, even with a smaller profit, the sugar-grower, by increased production, may find his business reasonably remunerative, if not as enriching as before. The sorghum-sugar grower will be injured or benefited with the growers of other kinds of sugar by these economic forces. Hence there should be no enmity between the grower of the sorghum, the sugar-beet, and the sugar-cane, but all should work in harmony for the general good.

It is true the present outlook is discouraging. But discouragement is not defeat. The time has now come for solid, energetic work. Science and practice must join improved agriculture, and all together can accomplish what neither alone would ever be able to achieve. It is not wise to promise too much, but this Bureau would fall short of its duty were it either to suppress the discouraging reports of this industry or fail to recognize the possibility of its success. The future depends on the persistence and wisdom of the advocates of sorghum. The problem they have to solve is a most difficult one, but its solution is not impossible.

It must be confessed finally that the chief object of this last series of experiments, viz, to place the industry where private capital would see its way clear to its extension over a large area has not been attained.

It is now seen that much of what has been done is useless, and were the work to be gone over again these necessary mistakes of a first attempt would be avoided. Time, labor, and money could be saved.

What encouragement is just is offered to those who are willing to take up this work here and extend it.

The great difficulties in the way of extracting the sugar from the cane have been removed. The fact that sorghum, in certain circumstances, becomes a fine-sugar producing plant has been incontestably established. A suitable soil and climate have been found for growing the crop and manufacturing the sugar. Remaining difficulties in the way of success have been fairly and candidly pointed out.

Since the present appropriation was made for continuing and concluding these experiments, I consider that my connection with the development of the industry has ended. I leave the work with only one regret, and that is that the future of the sorghum-sugar industry is still in doubt.

EXPERIMENTS WITH SUGAR-CANE.

On the 1st of October I received instructions from you to purchase a few tons of sugar-cane in Louisiana and make some experiments with it at Fort Scott.

The managers of the Daily City Item newspaper of New Orleans, having learned of your intention, made arrangements with the Texas Pacific Railroad to transport this cane from Louisiana to Fort Scott for $4 per ton. The general freight agent of the Mississippi Valley Railroad offered to deliver the cane on the same terms.

I requested Hon. Edward J. Gay to purchase the cane, which he kindly consented to do.

The cane was cut early in the season, viz, October 25 to 30, and was brought as quickly as possible to the factory.

PRELIMINARY TRIAL.

On November 2, three car-loads of cane having arrived, a preliminary trial was made.

The weight of cane used in this trial was 63.75 tons.

CUTTING-MACHINE.

The cutters which worked so poorly with sorghum did well with sugar cane, and no trouble whatever was experienced in producing chips suitable to diffusion and at the rate of six tons per hour.

CHIP ELEVATOR.

The same trouble was experienced with the elevator that we had had to contend with so long with sorghum, and to an increased extent. The chips being heavier than sorghum, easily overweighted the elevator and caused it to clog. Considerable delay was caused by these annoyances.

THE DIFFUSION.

It was found at once that the temperature used for the diffusion of sorghum, viz, 70° C., was entirely too low to effect the extraction of sugar from sugar-cane.

The temperature was gradually raised to 90° centigrade before a satisfactory extraction was obtained. The chips lying closer together in the cell caused the circulation of the liquid in the battery to take place more slowly. It was clearly evident that the pressure afforded by the feed-tank of the battery, viz, two-thirds of an atmosphere, is not great enough to work a battery rapidly when twelve cells are under pressure.

ANALYSES OF THE CANES WORKED.

Samples of chips were taken from each cell until twelve were filled. These samples were then passed through a small mill and the juice ob tained subjected to analysis.

The juices thus obtained had the following composition:

	Total solids.	Sucrose.	Glucose.
	Per cent.	Per cent.	Per cent.
First sample	14. 6	10. 52	2. 22
Second sample ...	13. 3	10. 10	1. 79
Third sample	14. 6	10. 89	2. 08
Fourth sample ...	14. 4	9. 82	2. 28
Fifth sample	14. 4	10. 04	2. 02
Means......	14. 26	10. 28	2. 08

WEIGHT OF DIFFUSION JUICE.

From each cell were drawn off 1,000 liters of juice, or 1,040 kilograms.

The number of cells filled with chips was 60; the weight of each cell of chips was 2,125 pounds; weight of juice drawn off from each cell was 2,280 pounds, or 163 pounds more than the weight of cane used.

ANALYSES OF DIFFUSION JUICE.

The samples were taken from each charge of juice drawn. When twelve were taken the mixture was analyzed:

	Total solids.	Sucrose.	Glucose.
	Per cent.	Per cent.	Per cent.
First sample	7. 2	5. 01	1. 15
Second sample ...	10. 4	7. 51	1. 48
Third sample	10. 8	7. 72	1. 56
Fourth sample ...	10. 8	7. 47	1. 69
Fifth sample	11. 3	7. 73	1. 77
Means	10. 1	7. 06	1. 53

EXHAUSTED CHIPS.

Four samples of exhausted chips were taken. The first one was from the first five cells only. No samples were taken from the next nine cells, and after that the samples were taken regularly as before. Following are the analyses:

	Total solids.	Sucrose.	Glucose.
	Per cent.	Per cent.	Per cent.
First sample	3. 5	2. 34	. 29
Second sample ...	2. 1	. 55	. 12
Third sample	1. 6	Lost.	Lost.
Fourth sample ...	1. 8	. 82	. 18
Means	2. 3	1. 24	. 20

The samples of carbonatatated and sulphured juices were not taken with regularity. Nevertheless I give below their analyses:

CARBONATATED JUICES.

	Total solids.	Sucrose.	Glucose.
	Per cent.	Per cent.	Per cent.
First sample	7.0	4.57	.84
Second sample ...	11.1	8.05	1.20
Third sample	11.5	7.76	1.30
Fourth sample ...	10.3	7.70	1.32
Means	9.98	7.02	1.17

SULPHURED JUICES.

	Total solids.	Sucrose.	Glucose.
	Per cent.	Per cent.	Per cent.
First sample	6.7	4.48	.86
Second sample ...	11.0	8.12	1.30
Third sample ...	11.3	8.20	1.35
Fourth sample ...	11.0	8.bo	1.36
Means	10.0	7.21	1.22

COMPOSITION OF SEMI-SIRUP FROM ABOVE JUICES.

	Per cent.
Total solids...	55.4
Sucrose	43.3
Glucose..	7.86

FIRST SUGARS MADE.

The *masse-cuite* was put in cars on November 4 and stood four days before commencing to dry it.

It yielded of first sugars...pounds..	6,888
Of second sugars..do....	495
Total first and second sugars ....do....	7,383
Sugar per ton...do....	115.8
Sugar on weight of cane ....per cent..	5.79

PER CENT. OF TOTAL SUCROSE OBTAINED.

The expressed juice contained 10.28 per cent. sucrose. Reckoning the juice at

90 per cent. of the weight of the cane, gives percentage sucrose in cane	9.25
Per cent. sugar obtained	5.79
Per cent. of total sugar obtained	62.6

ANALYSIS OF FIRST SUGARS.

	Per cent.
Moisture	.95
Ash..	.39
Glucose	1.05
Undetermined	.71
Sucrose	96.90

SECOND TRIAL.

On November 6, all the cane having arrived, the second trial was made. The experience of the first attempt had shown how the great loss of sugar in the chips, especially in the beginning, might be avoided. The second run was, therefore, made with an initial temperature of nearly 90° C. The quantity of juice withdrawn at each time was also increased by 100 liters.

Weight of cane used.—The weight of cane used in the second trial was 83.25 tons.

ANALYSES OF THE CANES.

The samples of chips were taken as described before:

	Total solids.	Sucrose.	Glucose.
	Per cent.	*Per cent.*	*Per cent.*
First sample	15.06	11.30	1.89
Second sample	14.68	10.86	1.62
Third sample	14.93	10.46	1.66
Fourth sample	13.47	10.43	1.89
Fifth sample	14.59	10.62	1.88
Sixth sample	13.55	10.05	1.75
Means	14.38	10.62	1.78

ANALYSES OF DIFFUSION JUICES.

The samples were taken as before described:

	Total solids.	Sucrose.	Glucose.
	Per cent.	*Per cent.*	*Per cent.*
First sample	10.11	7.33	1.18
Second sample	10.15	7.95	1.20
Third sample	10.08	7.15	1.17
Fourth sample	10.05	6.96	1.29
Fifth sample	9.83	7.03	1.29
Sixth sample	8.96	6.55	1.22
Means	9.86	7.16	1.23

EXHAUSTED CHIPS.

The samples were taken as described in the preliminary trial:

	Total solids.	Sucrose.	Glucose.
	Per cent.	*Per cent.*	*Per cent.*
First sample	1.56	.50	.12
Second sample	1.21	.38	.07
Third sample	1.11	.38	.10
Fourth sample	1.11	.37	.09
Fifth sample	1.06	.42	.10
Sixth sample	.77	.18	.05
Means	1.14	.37	.09

CARBONATATED JUICES.

The samples were taken in such a way as to represent the same body
of juice corresponding to the same numbered samples of diffusion juice.
Each carbonatation tank held three charges of diffusion juice. A meas-
ured sample after carbonatation was taken from each series of four
tanks.

	Total solids.	Sucrose.	Glucose.
	Per cent.	Per cent.	Per cent.
First sample..........	10. 11	7. 27	1. 09
Second sample	10. 25	7. 01	1. 14
Third sample.........	10. 14	7. 25	1. 11
Fourth sample........	9. 72	7. 00	1. 21
Fifth sample..........	9. 72	7. 10	1. 22
Sixth sample....	9, 55	6. 50	1. 12
Means..........	9. 92	7. 17	1. 15

SULPHURED JUICES.

The samples of sulphured juice were taken in a way to represent as
nearly as possible the same body of juice as indicated by the corresponding numbers under carbonatated juice. Since, however, the juices
after carbonatation had to fall into a receiving tank before being sent
to the filter presses, some mixing of the different bodies of juice was
unavoidable.

Thus the analyses below are not strictly comparable with the same
numbers under diffusion and carbonatated juices:

	Total solids.	Sucrose.	Glucose.
	Per cent.	Per cent.	Per cent.
First sample..........	9. 88	7. 68	1. 09
Second sample........	11. 12	8. 09	1. 14
Third sample.........	10. 35	7. 39	1. 23
Fourth sample........	9. 89	7. 02	1. 26
Fifth sample..........	10. 15	6. 93	1. 28
Sixth sample.........	9. 34	6 44	1. 17
Means..........	10. 12	7. 18	1. 20

SEMI-SIRUPS.

The semi-sirup from the above juices was put in two tanks. Samples
were taken from each tank:

	Total solids.	Sucrose.	Glucose.
	Per cent.	Per cent.	Per cent.
First sample..........	42. 9	32. 0	5. 95
Second sample	41. 9	30. 8	6. 45

The first sample represents the first third of the run, and the second
samples the second two-thirds.

FIRST SUGARS MADE.

The *masse-cuite* stood in cars two days.

On drying it yielded..pounds.. 11,185
The yield of "seconds" was ....do... 805

Total weight produced ....do... 11,990

Sugar per ton...do... 144
Sugar to weight of cane ....per cent.. 7.2

PER CENT. TOTAL SUGAR OBTAINED.

	Per cent.
The juice contained...	10.62
And the cane..	9.56
Percentage sucrose obtained..	75.3

COMPOSITION OF THE FIRST SUGARS.

The sample was taken from each barrel as it was filled. The samples were all mixed well together and placed in a tight bottle, which was not opened until the sample for analysis was taken. It is, therefore, as fair a sample of the product made as could possibly be obtained. It gave of—

	Per cent.
Moisture..	.73
Ash	.14
Glucose..	.52
Undetermined	.61
Sucrose..	98.00

Compare this result with the work on Magnolia plantation last year, as found in Bulletin No. 11, p. 26:

	Pounds.
Weight first sugars per ton..	119
Weight second sugars per ton	29.75

Total first and second	148.75

	Per cent.
Percentage obtained...	7.44
Sucrose in juice..	12.11
Sucrose in cane	10.90
Percentage obtained...	68.3
Sucrose in cane at Magnolia...	10.90
Sucrose in cane at Fort Scott...	9.56

Difference	1.34

The increase in the yield per ton at Magnolia, had the cane been worked by diffusion, would have been, therefore, 26.8 pounds.

The yield of seconds at Fort Scott was surprisingly low. The molasses as it came from the centrifugals was full of crystals. About one-third its volume of warm water was added to this molasses and the crystals all dissolved before boiling. This may have diminished the yield.

The "thirds" have been placed in cars and set away until next fall.

The "thirds" fill five wagons, each containing 23 cubic feet, or in all 125 cubic feet, weighing approximately 10,000 pounds. Of this amount, 6,189 pounds are from the second run.

	Pounds.
The total product, therefore, is, sugar	11,990
Thirds, *masse cuite*	6,189
Total	18,179

Or 218.3 pounds per ton of cane worked. This is nearly 11 per cent. of the weight of cane used.

But calculated on the original *masse cuite*, which filled 9 cars, there would have been $9 \times 23 = 207$ cubic feet, or 18,837 pounds $= 226$ pounds per ton, or 11.3 per cent.

But the method of reckoning the increased production which has just been used is not a fair one, since it rests on the assumption that the sucrose in each case is equally available. But a moment's consideration will show that this is not the case.

The term "available sugar" is not a precise one. It may have many interpretations. In France, for instance, the *rendement* is calculated by deducting from the total sucrose twice the glucose and from three to five times the ash. This is a good rule for beet sugar, but in cane-juice the ash, being mostly calcium salts, is far less melassigenic than that of the beet-juice, made up chiefly of potassium compounds.

Another method of calculating "available sugar" is to diminish the percentage of sucrose by the difference between it and all the other solids in solution. This method is apt, however, to give results too low. In this uncertainty the term "available sugar" should always be accompanied by an explanation of the manner of making the calculation.

The yield of sugar obtained at Fort Scott, being the highest ever got from sugar-cane, may be taken as the true amount of "available sugar" until some better yields are reported.

Notice, for a moment, the relation of this yield to the respective quantities of sucrose and glucose present:

	Per cent.
Sucrose in juice	10.62
Sucrose in cane	9.56
Yield of sucrose	7.20
Difference between sucrose in cane and yield	2.36
Glucose in juice	1.78
Glucose in cane	1.60

Ratio of per cent. of glucose to per cent. of sucrose lost 1.5 nearly.

It appears, therefore, that the rational way to calculate "available sugar" when the quantities of sucrose and glucose in the canes are known is to diminish the percentage of sucrose by one and a half times the glucose.

Applying this method we have the following results:

Sucrose in cane	per cent..	9.56
One and a half times glucose in cane	do....	2.40
Theoretical available sugar	do....	7.16
Pounds per ton		143.2
Pounds per ton obtained		144

Sucrose in cane	per cent..	10,90
One and a half times glucose in cane	do....	1.38
Theoretical available sugar	do....	9.52
Pounds per ton		194.4
Pounds per ton obtained		148.75
Difference	pounds..	41.65

This shows in the most convincing manner that by the process of diffusion and carbonatation the yield of sugar from sugar-cane can be increased fully 30 per cent. over the best milling and subsequent treatment of the juice which has ever been practiced in this or in any other country.

If this be true of the best milling, it is easy to estimate the increase over the average milling of Louisiana. It is not extravagant to suppose that this increase will be fully 40 per cent.

But the problem may also be approached in another way. It has just been shown what the product would have been had the Fort Scott process been applied at Magnolia. It may now be asked, "What would have been the yield had the Magnolia process been applied at Fort Scott?"

The process used at Magnolia produced 148.75 pounds sugar from cane in which the available sugar was 190.4 pounds. The percentage of available sugar obtained was

$$148.75 \times 100 \div 190.4 = 78.1 \text{ per cent.}$$

The available sugar in the cane at Fort Scott was 7.16 per cent. Multiply this by .78 and the product, 5.58 will be the yield of sugar which the Magnolia process would have given at Fort Scott, or 111.6 pounds per ton. Deduct this from the quantity obtained and the remainder will represent the increased yield, viz, 32.4 pounds. Thus in whatever way the calculation is made it is seen that the processes of diffusion and carbonatation give a largely increased yield.

Another important question which arises is this, "Does this increased yield come wholly from the increased extraction, or is it partly due to the method of purifying the juice?" I will try to give a rational answer to this question based on the data of the analyses and the respective *rendements* given by the two processes.

The percentage of extraction at Magnolia was 78. Reckoning the juice at 90 per cent., the loss in juice was 12 per cent. The percentage of juice, and consequently of sugar extracted, was 86.6 per cent. The mean loss of sugar in the chips at Fort Scott was .38 per cent., and the quantity of sugar present was 9.56. The percentage of extraction was therefore 96 per cent. The gain in extraction by diffusion is therefore 9.4 per cent. It is thus evident that the large gain in yield, as established at Fort Scott, cannot be due wholly to the increased extraction of the sugar. It must therefore be largely due to the processes of depuration employed.

The process of carbonatation tends to increase the yield of sugar in three ways:

(1) It diminishes the content of glucose. This diminution is small when the cold carbonatation as practised at Fort Scott is used; yet, to at least once and a half its extent, it increases the yield of crystallized sugar.

(2) By the careful use of the process of carbonatation there is scarcely any loss of sugar. The only place where there can be any loss at all is in the press cakes, and when the desucration of these is properly attended to the total loss is trifling. The wasteful process of "skimming" is entirely abolished, and the increased yield is due to no mean extent to this truly economical proceeding.

(3) In addition to the two causes of increase already noted, and which are not sufficient to produce the large *rendement* obtained, must be mentioned a third, the action of the excess of lime and its precipitation by carbonic acid on the substances in the juice, which are truly melassigenic. Fully half of the total increase which the experiments have demonstrated is due to this cause. It is true the coefficient of purity of the juice does not seem to be much affected by the process, but it is evident that the treatment to which the juice is subjected increases in a marked degree the ability of the sugar to crystallize. This fact is most abundantly illustrated by the results obtained.

Not only this but it is also evident that the proportion of first sugars to all others is largely increased by this method. This is a fact which may prove of considerable economic importance.

It thus appears that the yield of sugar would be greatly increased by the application of carbonatation to mill juices. Since a complete carbonatation outfit can be erected for about $4,000 it would be well if some planter or syndicate of planters should give the process a trial.

These facts are worthy of closer consideration, inasmuch as the process of carbonatation has been fiercely and maliciously assailed as one which destroys both sugar and molasses.

WEIGHT OF DIFFUSION JUICE COMPARED WITH WEIGHT OF CANE WORKED.

Number of cells filled, 84.3

Weight chips in each cell = 83.25 = 1.033 tons = 2,066 pounds.

Weight juice drawn from each cell of chips 1,100 liters. Specific gravity 1.04 = 2,516.8 pounds.   _1805_

The weight of normal juice in 2,000 pounds of cane is 1,859.4 pounds. The additional weight of water added by diffusion is 657.4 pounds. _11.1_

The percentage of increase over normal juice 657.4 ÷ 1,859.4 = 39.4 per cent. This increase represents what is often called the "dilution" of the juice. The quantity of water to be evaporated to produce a given quantity of sugar is, therefore, 39.4 per cent. greater for such a diffusion than for a normal mill juice. In practice this amount could easily be reduced to 25 per cent.

COMPOSITION OF PRESS CAKE.

The defecation and filtration of the juice from 83.25 tons of cane gave 197 press cakes.

The mean weight of these cakes was 24 pounds each, and the total weight 4,728 pounds. A sample of the cake taken directly from the press and dried contained of moisture 45.37 per cent. The total weight of dry matter obtained in the press cakes was, therefore, 2,582.9 pounds.

Analyses of the dried cake gave the following results:

	Per cent.
Albuminoids	9,585
Sucrose	Trace.
Glucose	Trace.
Other organic matter	17.45

QUANTITY OF LIME USED.

As is seen under sorghum experiments it required 1.5 per cent. lime to produce a good filtration.

I felt sure that the juice from the sugar-cane would not require as great a quantity. At the preliminary trial 1 per cent. of lime was used and the cakes formed were perfect, firm, and hard.

In the second run only .75 per cent. of lime was used, and the cakes were equally as good. There is little occasion for using less lime than this, for with this quantity the carbonatations were easily finished in fifteen to twenty minutes.

COEFFICIENT OF PURITY IN SECOND TRIAL.

	Per cent.
Of the mill juices the coefficient was	73.8
Of the diffusion juices the coefficient was	72.6
Of the carbonatated juices the coefficient was	72.3
Of the sulphured juices the coefficient was	70.9
Of the first semi-sirup the coefficient was	74.6
Of the second semi-sirup the coefficient was	73.5

In both trials it was seen that the coefficient of purity was increased during the process of evaporation. This was, doubtless, caused by the precipitation of some of the lime salts held in solution by the juices.

a few days before the experiment was made, but it was still black and putrid, emitting a nauseating stench.

The strike-pan used was quite unsuitable for boiling to grain. Its base was once the bottom of a much smaller pan, and a shelf several inches deep had been added to support the enlarged top. All the large steam-coils were above this shelf, and it took eight hours to bring the contents of the pan above this point. We had no sugar-boiler, but my assistant, Mr. G. L. Spencer, took charge of the pan and did remarkably well.

The sugar dried slowly in the centrifugals. These were not well set and could not be run at a very high speed on account of shaking.

It took nearly forty-eight hours with three machines to dry the sugar from the 83.25 tons.

This difficulty in drying was due either—

(1) To the process of diffusion; (2) to the process of carbonatation; (3) to the fine grain produced in boiling; (4) or to the poor quality of the cane.

Which one of these causes was most potent only future experiments will decide. I am not wise enough to place it, as has already been done by some premature critics, on one of them alone.

It seems most reasonable to suppose, however, that the poor quality of the cane and the extreme fineness of the crystals were the chief causes of the difficulty mentioned. The process of carbonatation has been practiced for ten years in Java on mill juices and no complaint has ever been heard of difficulty in purging the sugar. With the fresh, ripe canes of Louisiana worked promptly as they come from the field, and with the juice in the hands of an experienced sugar-boiler, I do not believe this difficulty would be encountered.

With the improvements in the process of carbonatation already pointed out in the discussion of the experiments with sorghum even better results may be expected.

Erratum :

In lieu of article on "weight of Diffusion juice compared with weight of cane worked" pp. 53 and 54, Bul. No. 14, read as follows:

Number of cells filled 83.

Weight chips in each cell $= 83.25 \div 83 = 1.003$ tons $= 2006$ pounds.

Normal weight of juice in 2006 pounds of cane 1805 pounds. Additional weight of water added by diffusion 711.8 pounds. Percentage of increase over normal juice 711.8 $\times$ 100 $\div$ 1805 $= 39.4$

press and dried contained of moisture 45.37 per cent. The total weight of dry matter obtained in the press cakes was, therefore, 2,582.9 pounds.

Analyses of the dried cake gave the following results:

	Per cent.
Albuminoids	9,585
Sucrose	Trace.
Glucose	Trace.
Other organic matter	17. 45

QUANTITY OF LIME USED.

As is seen under sorghum experiments it required 1.5 per cent. lime to produce a good filtration.

I felt sure that the juice from the sugar-cane would not require as great a quantity. At the preliminary trial 1 per cent. of lime was used and the cakes formed were perfect, firm, and hard.

In the second run only .75 per cent. of lime was used, and the cakes were equally as good. There is little occasion for using less lime than this, for with this quantity the carbonatations were easily finished in fifteen to twenty minutes.

COEFFICIENT OF PURITY IN SECOND TRIAL.

	Per cent.
Of the mill juices the coefficient was	73. 8
Of the diffusion juices the coefficient was	72. 6
Of the carbonatated juices the coefficient was	72. 3
Of the sulphured juices the coefficient was	70. 9
Of the first semi-sirup the coefficient was	74. 6
Of the second semi-sirup the coefficient was	73. 5

In both trials it was seen that the coefficient of purity was increased during the process of evaporation. This was, doubtless, caused by the precipitation of some of the lime salts held in solution by the juices.

DEGREE OF EXTRACTION BY EXPERIMENTAL MILL.

	Fresh chips per cent. juice obtained.	Exhausted chips per cent. water extracted.
First sample......	54. 64	63. 73
Second sample ..	58. 88	62. 68
Third sample.....	57. 61	63. 39
Fourth sample....	55. 85	62. 01
Fifth sample	60. 00	69. 65
Sixth sample.....	51. 48	60. 83
Mean........	56. 41	63. 72

DIFFICULTIES ENCOUNTERED.

A number of unfavorable conditions was encountered during the prosecution of the experiments. The water supply was from a stagnant pond. The water had been greatly improved by the application of lime a few days before the experiment was made, but it was still black and putrid, emitting a nauseating stench.

The strike-pan used was quite unsuitable for boiling to grain. Its base was once the bottom of a much smaller pan, and a shelf several inches deep had been added to support the enlarged top. All the large steam-coils were above this shelf, and it took eight hours to bring the contents of the pan above this point. We had no sugar-boiler, but my assistant, Mr. G. L. Spencer, took charge of the pan and did remarkably well.

The sugar dried slowly in the centrifugals. These were not well set and could not be run at a very high speed on account of shaking.

It took nearly forty-eight hours with three machines to dry the sugar from the 83.25 tons.

This difficulty in drying was due either—

(1) To the process of diffusion; (2) to the process of carbonatation; (3) to the fine grain produced in boiling; (4) or to the poor quality of the cane.

Which one of these causes was most potent only future experiments will decide. I am not wise enough to place it, as has already been done by some premature critics, on one of them alone.

It seems most reasonable to suppose, however, that the poor quality of the cane and the extreme fineness of the crystals were the chief causes of the difficulty mentioned. The process of carbonatation has been practiced for ten years in Java on mill juices and no complaint has ever been heard of difficulty in purging the sugar. With the fresh, ripe canes of Louisiana worked promptly as they come from the field, and with the juice in the hands of an experienced sugar-boiler, I do not believe this difficulty would be encountered.

With the improvements in the process of carbonatation already pointed out in the discussion of the experiments with sorghum even better results may be expected.

The disposition of the exhausted chips is a question of great economic importance. Three uses appear to be possible : (1) For paper stock ; (2) for manure ; (3) for fuel.

A good article of both wrapping and print paper can be made of the fiber of the cane. The economic discussion of this use, however, can only be properly given by a paper-maker.

The value of the bagasse for a manure is undoubtedly great. This problem has already been discussed in Bulletin No. 8, page 46.

By referring to the table of analyses of the chips it will be seen that with a small hand-mill 63.72 per cent. of water was extracted from the exhausted chips; on the same mill the percentage of extraction of the fresh chips was only 56.31 per cent. Thus in similar conditions the percentage of extraction with a given mill will be 7.31 per cent. higher for exhausted chips than for fresh canes. A mill, therefore, which will give a 78 per cent. extraction with cane will give 85 per cent. with exhausted chips.

The exhausted chips contained 90 per cent. water. Of this quantity 63.72 per cent. were extracted, leaving 26.28 per cent. water to 10 fiber. A given quantity of the bagasse, therefore, contained 72.2 per cent. water and 27.8 per cent. fiber. A mill which would give 80 per cent. extraction with the exhaused chips would furnish a bagasse composed of equal parts of water and fiber and this would prove a most excellent fuel.

The power required to drive such a mill would only be about one-third as great as for the same weight of cane.

The attempts to dry cane chips on the presses used for beet cuttings have proved failures, but the experiments made at Fort Scott show that a properly arranged mill will solve this problem at once.

It must be remembered, however, that even if the exhausted chips be made as dry as ordinary mill bagasse they will not afford so much fuel. They contain little but the fiber of the cane, while mill bagasse still holds large quantities of sugar, which itself is a most excellent fuel.

The loss of the bagasse as a fuel has been the principal objection to the introduction of diffusion into tropical sugar districts.

It now remains to continue these experiments at some favorable station in Louisiana. Such a station should be provided with a first-class double or triple effect and other apparatus for evaporating the juice and separating the sugar.

It should also be a station purely experimental. The attempt to carry on experiments and manufacture a large crop of cane at the same time would only end in the disastrous manner, economically considered, of the sorghum work just concluded at Fort Scott.

These experiments can only be successful at a station where perfect freedom of action and plenty of time are at the director's command.

It is the proper province of the Department to demonstrate in Lou. isiana just how much increase in sugar yield can be produced by the application of the methods named in the act making the appropriations. This done, and all the processes for doing it accurately pointed out and logically discussed, it will not be difficult for the intelligent planter to determine the economic value of the new methods.

To this task should be brought a careful study of the chemical problems involved, and the best apparatus which this country or Europe can afford. From this task should be eliminated all prejudices for or against any particular process, and especially all tendency to misrepre. sent or misinterpret facts.

At least the Department will be able in subsequent experiments to show the Southern sugar-raiser whether the promises which these preliminary experiments have made shall really be performed, or whether the practice of the process of diffusion for sugar-cane is a mistake and the prospects it has offered of aiding the sugar industry a delusion.

It is certain that with the fierce rivalry between the European beet and the tropical cane industry, producing an enormous surplus of sugar and sending the prices down almost below the cost of production, the indigenous sugar-cane industry of this country will languish unless the Department of Agriculture be able to lead it into a life of renewed vigor.

INDEX.

63

www.ingramcontent.com/pod-product-compliance
Lightning Source LLC
Chambersburg PA
CBHW022007190326
41519CB00010B/1413